International Conference on CRISPR Technologies 2019

Wurzburg, Germany
16-18 September 2019

ISBN: 978-1-7138-0731-5

Printed from e-media with permission by:

Curran Associates, Inc.
57 Morehouse Lane
Red Hook, NY 12571

Printed with permission by Curran Associates, Inc. (2020)

For permission requests, please contact AIChE
at the address below.

AIChE
120 Wall Street, FL 23
New York, NY 10005-4020

Phone: (800) 242-4363
Fax: (203) 775-5177

www.aiche.org

Additional copies of this publication are available from:

Curran Associates, Inc.
57 Morehouse Lane
Red Hook, NY 12571 USA
Phone: 845-758-0400
Fax: 845-758-2633
Email: curran@proceedings.com
Web: www.proceedings.com

TABLE OF CONTENTS

MEDICAL APPLICATIONS

AGRICULTURAL APPLICATIONS

SOCIETAL IMPACTS

BACTERIAL APPLICATIONS

POSTER SESSION

REGISTERED POSTERS

A Positive, Growth-Based PAM Screen Reveals Non-Canonical Motifs Recognized By the S. Pyogenes Cas9

CRISPR TECHNOLOGIES CONFERENCE

Conference: International Conference on CRISPR Technologies

Year: 2019

Proceeding: 3rd International Conference on CRISPR Technologies

Group: General Submissions

Session: CRISPR Cas-Biology

Time: Monday, September 16, 2019 - 10:05am-10:30am

Authors:

Collias, D., North Carolina State University
Leenay, R. T., North Carolina State University
Slotkowski, R. A., North Carolina State University
Collins, S., North Carolina State University
McGirr, B., North Carolina State University
Jacobsen, T., North Carolina State University
Beisel, C., Helmholtz Institute for RNA-based Infection Research

CRISPR technologies have overwhelmingly relied on the ubiquitous and extensively studied Cas9 from *Streptococcus pyogenes* (SpCas9). One well-defined feature of SpCas9 is its recognition of the canonical NGG and less-preferred NAG and NGA protospacer-adjacent motifs (PAMs) as the complete set used for target selection and off-target predictions. Here, we report that SpCas9 recognizes additional motifs at levels similar to that of the less-preferred PAMs. These motifs were identified using a positive, growth-based screen in *E. coli* that surprisingly enriched N(A/C/T)GG sequences over the consensus NGG. The enriched N(A/C/T)GG motif effectively represents an extra N between the target and consensus NGG. DNA binding and cleavage assays in bacteria validated the screenâ€™s enriched sequences while indel formation and subsequent GFP disruption validated the N(C/T)GG motif in mammalian cells. Within these experiments, we found that the first two PAM nucleotides contributed to the overall targeting activity. We also discovered that high-fidelity SpCas9â€™s recognized the N(C/T)GG motif, while the editing location for SpCas9-derived base editors could be shifted by selecting (C/T)GG or N(C/T)GG PAMs. Finally, we found that these motifs were present but overlooked in high-throughput spacer acquisition and off-target assays. SpCas9 therefore exhibits greater targeting flexibility than previously recognized, with implications for precise editing, off-target predictions, and CRISPR-based immunity.

DNA Uptake into CRISPR Loci

Conference: International Conference on CRISPR Technologies

Year: 2019

Proceeding: 3rd International Conference on CRISPR Technologies

Group: General Submissions

Session: CRISPR Cas-Biology

Time: Monday, September 16, 2019 - 9:40am-10:05am

Authors:

Terns, M., University of Georgia

Acquiring foreign spacer DNA into the CRISPR locus is an essential primary step of the CRISPR-Cas pathway in prokaryotes for developing host immunity to mobile genetic elements. Pyrococcus furiosus, a hyperthermophilic archaeaon, utilizes one set of adaptation proteins (Cas1, Cas2, Cas4-1 and Cas4-2) for spacer DNA uptake into each of its seven CRISPR loci. We have shown that the 200 different crRNAs naturally encoded by the seven CRISPR arrays are each incorporated into three functional crRNP effector complexes, Type I-A (Csa), Type I-G (Cst), and Type III-B (Cmr), to destroy matching plasmid DNA through either DNA- or RNA-targeting mechanisms. Recently, we found that Pyrococcus furiosus actively takes up spacers into each of its seven CRISPR loci under normal growth conditions and without the need for over-expressing adaptation proteins.

Here, we investigate spacer integration in vitro using proteins from Pyrococcus furiosus and demonstrate that Cas1 and Cas2 are sufficient to accurately integrate spacers into a minimal CRISPR locus. Using high-throughput sequencing, we identified high frequency spacer integration occurring at the same CRISPR repeat border sites utilized in vivo, as well as at several non-CRISPR plasmid sequences which share features with repeats. Analysis of non-CRISPR integration sites revealed that Cas1 and Cas2 are directed to catalyze full-site spacer integration at specific DNA stretches where guanines and/or cytosines are 30 base pairs apart and the intervening sequence harbors several positionally conserved bases. Moreover, assaying a series of CRISPR repeat mutations, followed by sequencing of the integration products, revealed that the specificity of integration is primarily directed by sequences at the leader-repeat junction as well as an adenine-rich sequence block in the mid-repeat. Together, our results indicate that P. furiosus Cas1 and Cas2 recognize multiple sequence features distributed over a 30 base pair DNA region for accurate spacer integration at the CRISPR repeat.

Functional Metagenomic Identification of Anti-CRISPR Proteins

Conference: International Conference on CRISPR Technologies

Year: 2019

Proceeding: 3rd International Conference on CRISPR Technologies

Group: General Submissions

Session: CRISPR Cas-Biology

Time: Monday, September 16, 2019 - 11:20am-11:45am

Authors:

Sommer, M., Technical University of Denmark

CRISPR-technologies represent effective biotechnological tools that are broadly impacting life sciences. Yet, the control over CRISPR-technologies remains poorly understood. Inhibitors of CRISPR-Cas systems offer opportunities to modulate its activity, yet only few such systems are available. Using a synthetic genetic circuit we developed a high-throughput approach to discover anti-CRISPR genes from metagenomic libraries based on their functional activity rather than sequence homology or genetic context. We identified DNA fragments from soil, animal and human metagenomes that circumventStreptococcus pyogenes Cas9 activity in our selection strain. Further characterization of a subset of these hits validate the activity of novel anti-CRISPRs. Notably, homologues of these anti-CRISPRs found in diverse hosts from seven different phyla have high sequence identity indicating recent horizontal gene transfer. Collectively our data demonstrates that anti-CRISPRs against type II-A CRISPR-Cas systems are widely distributed across bacterial phyla suggestive of a more complex ecological role than previously appreciated.

Harnessing bacterial immune system for targeted genome engineering

CRISPR
TECHNOLOGIES
CONFERENCE

Conference: International Conference on CRISPR Technologies

Year: 2019

Proceeding: 3rd International Conference on CRISPR Technologies

Group: General Submissions

Session: CRISPR Cas-Biology

Time: Monday, September 16, 2019 - 9:15am-9:40am

Authors:

Siksnys, V., Vilnius University

Bacteriophages infect bacteria in order to reproduce and often pose a lethal threat to bacteria. To cope with onslaughts of phages bacteria evolved multiple defense barriers that interfere with nearly every step of virus life cycle. Taken together, these defense barriers constitute a primitive immune system that protects bacteria against invading viruses. In prokaryotes restriction-modification system acts as an innate immune system that uses methyl-tag to discriminate between self and non-self DNA, while CRISPR-Cas system functions as an adaptive arm that memorizes the invader by inserting fragments of the foreign DNA into the host chromosome. Class 2 CRISPR-Cas systems based on the single effector protein, as exemplified by Cas9, rely on the protospacer adjacent motif (PAM) sequence to discriminate between host and alien DNA. Deciphering the molecular mechanisms underlying the CRISPR-Cas immunity paved the way for development of novel tools for targeted genome engineering. Class 2 effector nucleases that provide CRISPR-Cas immunity are abundant in prokaryotes. To explore this largely uncharacterized diversity of the Class2 nucleases for genome editing applications, we established a phylogeny-guided bioinformatic approach and developed biochemical screens for the rapid identification and characterization of the PAM and guide RNA requirements of new nucleases. This approach permitted the rapid characterization of many Cas9 orthologs with diverse PAM sequence requirements. Next, using the same biochemical screen we detected PAM and gRNA requirements for several miniature CRISPR-associated proteins. Taken together, our results demonstrate that Class 2 effector nucleases provide a rich source of biochemical and biophysical diversity that may be beneficial for genome editing.

RNA-targeting by Cas13a effector and DNA-proficient spacer acquisition in Type VI-A CRISPR-Cas system: How to encounter enemy, not die and prevent infection spreading

Conference: International Conference on CRISPR Technologies

Year: 2019

Proceeding: 3rd International Conference on CRISPR Technologies

Group: General Submissions

Session: CRISPR Cas-Biology

Time: Monday, September 16, 2019 - 10:55am-11:20am

Authors:

Semenova, E., Rutgers University

Among evolutionary and mechanistically diverse CRISPR-Cas systems, Type VI systems are unique as they exclusively target RNA. Type VI single-protein effector Cas13 exhibits a dual activity: Cas13 is responsible for crRNA biogenesis as well as for target RNA destruction. Significant progress has been made in understanding the mechanism of Type VI crRNA biogenesis. In contrast, little is known about interference and adaptation in these systems. Interference by Cas13 has been mostly studied using in vitro assays, which revealed that target RNA recognition activates collateral damage of non-complementary RNAs. Cas13 collateral RNase activity has been adopted for developing of highly sensitive diagnostic platforms. Observation of growth defect exhibited by cells undergoing Type VI interference with non-essential targets led to a hypothesis of programmed cell death/dormancy induced by Cas13 collateral damage activity.

To investigate a mechanism of cell growth inhibition we study Cas13a system from Leptotrichia shahii transplanted into Escherichia coli cells. Live microscopy and pulse-labeling experiments showed a cessation of cell division probably due to inhibition of protein synthesis upon Cas13a RNA-targeting. We suggest that Cas13a system likely switches to cell dormancy rather than programmed cell death. RNA sequencing and primer extension assay revealed numerous RNA cleavage sites in mRNAs, rRNAs, and tRNAs. The analysis also detected the Cas13a-mediated activation of type II toxin-antitoxin systems. Using strains deficient in toxin RNases we aim to explore a predominate mechanism for induction of dormancy state.

Another intriguing questions are how do the cells that carry type VI CRISPR-Cas system can memorize phage infection and how can spacer acquisition be compatible with induced cell dormancy. I will present our results on adaptation in Type VI-A system and discuss why cell dormancy would be more beneficial for the population than cell death to retain immunity status against invaders.

Transcriptional and post-transcriptional regulation of type I and III CRISPR-Cas activity

CRISPR TECHNOLOGIES CONFERENCE

Conference: International Conference on CRISPR Technologies

Year: 2019

Proceeding: 3rd International Conference on CRISPR Technologies

Group: General Submissions

Session: CRISPR Cas-Biology

Time: Monday, September 16, 2019 - 10:30am-10:55am

Authors:

Fineran, P., University of Otago

Although CRISPR-Cas systems benefit their hosts by providing phage resistance, other costs, such as autoimmunity, may occur. One way that bacteria might maximize the benefits of their CRISPR-Cas systems is via regulation â€“ e.g. by upregulating the systems when there is a high risk of infection or a phage epidemic. By developing a new method and using existing approaches, we have identified a suite of regulators that influence Class 1 (type I and III) expression and activity. In addition to cell-density dependent regulation (quorum sensing), we have identified both transcriptional and post-transcriptional regulators of CRISPR-Cas activity.

CRISPR Switches: Non Sequence-Constrained RNA-Sensing Grnas

CRISPR TECHNOLOGIES CONFERENCE

Conference: International Conference on CRISPR Technologies

Year: 2019

Proceeding: 3rd International Conference on CRISPR Technologies

Group: General Submissions

Session: Foundational Technologies

Time: Monday, September 16, 2019 - 3:40pm-4:05pm

Authors:

Rostain, W., Institut Franaois Jacob, CEA

We all like CRISPR, for its uses in genome engineering or for the panoply of engineered Cas proteins that have since transformed CRISPR into an RNA-guided platform with diverse uses. The presence of an RNA moiety in CRISPR-Cas systems offers the opportunity to link the field of CRISPR with that of RNA engineering, which uses secondary structure modelling to create artificial RNA sequences that can carry out logical operations. We have designed engineered Cas12a gRNAs that are rendered constitutively inactive by the addition of artificial structural elements, but can be switched back on upon the addition of a second trigger RNA. Making use of a cell-free transcription-translation system to characterise the engineered gRNAs, we were able to explore a diversity of designs for these CRISPR switches, shedding light on the importance and optimal design of the various structural elements that compose them. This permitted the creation of RNA-sensing CRISPR gRNAs that can be activated by artificial or natural RNA sequences. The switches are not sequence-constrained by the Cas12a handle or by the guide sequence, so could theoretically be used to detect any sequence without additional control elements. They can be used to control the activity of Cas proteins*in vitro* and *in vivo,* providing an additional layer of control and permitting the activation of Cas proteins upon the expression of an exogeneous or endogenous RNA.

CRISPR-Cas9 Mediated Endogenous RNA Targeting in the Foodborne-Pathogen Campylobacter Jejuni

CRISPR TECHNOLOGIES CONFERENCE

Conference: International Conference on CRISPR Technologies

Year: 2019

Proceeding: 3rd International Conference on CRISPR Technologies

Group: General Submissions

Session: Foundational Technologies

Time: Monday, September 16, 2019 - 2:25pm-2:50pm

Authors:

Sharma, S., Institute of Molecular Infection Biology (IMIB), University of Wurzburg
Beisel, C. L., Helmholtz Institute for RNA-based Infection Research
Sharma, C., Institute of Molecular Infection Biology (IMIB), University of Wurzburg
Bischler, T., Institute of Molecular Infection Biology (IMIB), University of Wurzburg

The CRISPR (clustered regularly interspaced short palindromic repeats)-associated protein Cas9 is a sequence-specific nuclease, which typically cleaves double-stranded target DNA with the help of CRISPR-RNAs (crRNAs) guides and a trans-activating CRISPR RNA (tracrRNA) (1). In a global transcriptome study, we observed that the CRISPR locus of the food-borne pathogen *Campylobacter jejuni* is one of the most highly transcribed regions in the genome (Dugar 2013 Plos Genetics); however, its potential role beyond genome defense is unclear. Recently, we have uncovered that the Cas9 from *Campylobacter jejuni* (CjCas9) can also bind and cleave endogenous mRNAs in a crRNA-dependent manner (2). This discovery raises questions on whether Cas9 could play a role in mediating endogenous gene regulation and virulence or whether RNA targeting by CjCas9 is just an off-target effect. To delineate the potential role of endogenous RNA targeting via CRISPR-Cas9, we investigate the regulation of tracrRNA and crRNAs in different strains of *C. jejuni* and aim to identify conditions that might affect RNA targeting in *C. jejuni* via CRISPR locus expression. Also, it is not known whether transcript stability and processing of crRNAs and tracrRNA play any role in maintaining crRNA-mRNA interactions. Using transcript stability assays and RNase-deletion mutant strains, we aim to identify factors that control their stability and interactions with Cas9. Using RIP-seq combined with computational analysis, we are investigating endogenous RNA targets and their cleavage by Cas9 in different *C. jejuni* strains with varying number of crRNAs. Identifying and understanding the functions of the CRISPR locus, its transcribed products and potential target RNAs, will provide insights into the newly discovered world of CRISPR-Cas9 mediated endogenous RNA targeting and its possible role in growth and virulence of *C. jejuni*.

Engineering Programmable Tools to Target the Transcriptome

CRISPR
TECHNOLOGIES
CONFERENCE

Conference: International Conference on CRISPR Technologies

Year: 2019

Proceeding: 3rd International Conference on CRISPR Technologies

Group: General Submissions

Session: Foundational Technologies

Time: Monday, September 16, 2019 - 2:00pm-2:25pm

Authors:

Stafforst, T., University of Tubingen

Programmable RNA-guided tools for gene editing attracted a lot of interest during the last years. However, there are safety concerns related to off-target edits, and immunogenicity, and gene editing fails when the introduced mutation is lethal or genetically compensated. For the last years, our lab has been pioneering several alternative strategies that are based on site-directed A-to-I editing of specific transcripts.

Specifically, we developed the SNAP-ADARs approach that applies a unique and novel assembly strategy to generate artificial RNA-guided riboproteins. Recently, we demonstrated the simultaneous A-to-I editing of several endogenous transcripts, with high efficiency (up to 90%), high potency, sufficient duration, and high precision [1]. Yet unpublished, we will report on improvements of the approach and their application on the manipulation of signaling proteins. We aim to point out that the SNAP-tag approach is a simple and convenient general RNA-targeting approach, orthogonal to current Cas approaches and may allow for the targeted writing and erasing of further epitranscriptomic marks.

Furthermore, we develop genetically encodable guideRNAs [2] and chemically stabilized oligonucleotides [3] that enable the recruitment of endogenous ADARs for site-directed RNA editing with high potential for clinical translation.

[1] Vogel et al., Efficient and Precise Editing of Endogenous Transcripts with SNAP-tagged ADARs. Nat. Meth. 15, 535 (2018).

[2] Wettengel, J., Reautschnig, J., Geisler, S., Kahle, P. J., Stafforst, T. Harnessing human ADAR2 for RNA repair â€" Recoding a PINK1 mutation rescues mitophagy. Nucl. Acids Res. 45, 2797-2808 (2017).

[3] T. Merkle, S. Merz, P. Reautschnig, A. Blaha, Q. Li, P. Vogel, J. Wettengel, J. B. Li, T. Stafforst. Precise RNA editing by recruiting endogenous ADARs with antisense oligonucleotides. Nature Biotech. 37, 133-38 (2019).

Massively Multiplexed Nucleic Acid Detection with Cas13

Conference: International Conference on CRISPR Technologies

Year: 2019

Proceeding: 3rd International Conference on CRISPR Technologies

Group: General Submissions

Session: Foundational Technologies

Time: Monday, September 16, 2019 - 3:15pm-3:40pm

Authors:

Myhrvold, C., Harvard University
Ackerman, C. M., Broad Institute
Thakku, S. G., Broad Institute of Harvard and MIT
Freije, C. A., Broad Institute of Harvard and MIT
Metsky, H., Broad Institute of Harvard and MIT
Yang, D., Broad Institute of Harvard and MIT
Kehe, J., Massachusetts Institute of Technology
Carter, A., Broad Institute of Harvard and MIT
Kulesa, A., Broad Institute of Harvard and MIT
Blainey, P. C., Massachusetts Institute of Technology
Sabeti, P., Harvard T.H. Chan School of Public Health
Hung, D., Broad Institute of Harvard and MIT

The overwhelming majority of pathogens that circulate in human and animal populations go undetected, undermining patient care and hindering outbreak preparedness. This is largely due to a lack of inexpensive, scalable technologies for pathogen detection and surveillance. CRISPR-based nucleic acid detection methods such as SHERLOCK and DETECTR are emerging solutions to the problem of pathogen surveillance, but are limited to testing 1-4 targets per assay. Here, we introduce Combinatorial Arrayed Reactions for Multiplexed Evaluation of Nucleic acids (CARMEN), a technology that uses self-organization and miniaturization to enable highly multiplexed nucleic acid detection. Using CARMEN, we demonstrate parallelized CRISPR-Cas13 detection with up to 5,000 crRNA-target pairs tested in a single assay, increasing multiplexing and throughput while simultaneously decreasing the reagent cost per test by >300-fold. Using CARMEN-Cas13, we designed and extensively tested a 169-plex assay that simultaneously differentiates all human-associated viruses with â‰¥10 available genome sequences. CARMEN-Cas13 also enables comprehensive subtyping of influenza A strains and multiplexed identification of dozens of HIV drug-resistance mutations. Highly multiplexed CRISPR-based nucleic acid detection can catalyze rapid molecular diagnosis and characterization of wide-ranging pathogens, greatly benefiting patients and public health.

Programmable Inhibition and Detection of RNA Viruses Using Cas13

CRISPR TECHNOLOGIES CONFERENCE

Conference: International Conference on CRISPR Technologies

Year: 2019

Proceeding: 3rd International Conference on CRISPR Technologies

Group: General Submissions

Session: Foundational Technologies

Time: Monday, September 16, 2019 - 2:50pm-3:15pm

Authors:

Freije, C. A., Broad Institute of Harvard and MIT
Myhrvold, C., Harvard University
Boehm, C. K., Broad Institute
Lin, A. E., Broad Institute
Carter, A., Broad Institute of Harvard and MIT
Metsky, H., Broad Institute of Harvard and MIT
Luo, C. Y., Broad Institute
Abudayyeh, O., Broad Institute/MIT
Gootenberg, J., Broad Institute
Yozwiak, N. L., Broad Institute
Zhang, F., Broad Institute of MIT and Harvard/McGovern Institute for Brain Research/Massachusetts Institute of Technology
Sabeti, P., Harvard T.H. Chan School of Public Health

Two-thirds of known viruses have RNA genomes, and numerous of these cause human disease yet are lacking virus-specific antivirals or vaccines. Antiviral development is limited by our knowledge of viral and host protein biology and is compounded by virusesâ€™ ability to evolve resistance to existing antivirals. The RNA-guided, RNA-targeting CRISPR-Cas effector Cas13 could enable a new class of CRISPR-based antivirals. Moreover, Cas13 is highly programmable as it can cleave any RNA complementary to its CRISPR RNA (crRNA). Here, we demonstrate that Cas13 can potently reduce viral RNA >15-fold and infectivity >300-fold for two distinct RNA viruses: lymphocytic choriomeningitis virus (LCMV), a model arenavirus for studying the deadly Lassa fever virus, and influenza A virus. We perform a tiling screen along the LCMV genome and identify over 100 cRNAs that reduce LCMV levels by â‰¥2-fold and tens of crRNAs that reduce levels by â‰¥10-fold. Analysis of the target sites of these crRNAs reveals targeting â€œhotspotsâ€ along the LCMV genome and helps define a set of crRNA design principles for targeting viral RNA. We demonstrate that targeting efficiency can be enhanced through the use of alternative Cas13 orthologs and multiplexing. Additionally, we develop CARVER (Cas13-Assisted Restriction of Viral Expression and Readout), an end-to-end platform for Cas13-based detection of viral RNA in Cas13-treated samples using SHERLOCK (Specific High-sensitivity Enzymatic Reporter UnLOCKing). We showcase the remarkable promise of CRISPR-Cas13 to enable both diagnostic and antiviral technologies, which could have widespread implications for future diagnosis and treatment of human infectious disease.

Transcriptional recording by CRISPR spacer acquisition from RNA

Conference: International Conference on CRISPR Technologies

Year: 2019

Proceeding: 3rd International Conference on CRISPR Technologies

Group: General Submissions

Session: Foundational Technologies

Time: Monday, September 16, 2019 - 4:30pm-4:55pm

Authors:

Schmidt, F., ETH Zurich
Platt, R., ETH Zurich

RNA sequencing is unmatched in its potential to characterize cell types, cell states, and the influence of perturbations on phenotypes. The drop in DNA sequencing has fueled progress in this area making it possible to evaluate diverse biological phenomenon at scale. However, the ability to interrogate dynamic cellular phenotypes, rather than a single snapshot in time, have been hampered by a lack of non-destructive methodologies. Moreover, RNA sequencing experiments reflect the current state of a cell, providing no information on its history. One promising solution is to store biological information in DNA, which is an ideal medium for encoding, storing, and propagating information. However, the ability to transform transcriptional information into DNA in living cells over time has been impossible due to a lack of molecular tools that facilitate recording (capture/encoding/storage) of RNA into the DNA. In my talk I will describe the development of Record-seq, a method leveraging CRISPR spacer acquisition from RNA to continuously record and permanently store transcriptomes inside populations of living cells, enabling the reconstruction of cellular histories.

Transposon-encoded CRISPR-Cas systems facilitate programmable DNA integration

Conference: International Conference on CRISPR Technologies

Year: 2019

Proceeding: 3rd International Conference on CRISPR Technologies

Group: General Submissions

Session: Foundational Technologies

Time: Monday, September 16, 2019 - 4:05pm-4:30pm

Authors:

Klompe, S., Caribou Biosciences

Transposon-encoded CRISPR-Cas systems facilitate programmable DNA integration

Ccexo: A New Tool for the Enhanced Genome Editing

CRISPR TECHNOLOGIES CONFERENCE

Conference: International Conference on CRISPR Technologies

Year: 2019

Proceeding: 3rd International Conference on CRISPR Technologies

Group: General Submissions

Session: Medical Applications

Time: Tuesday, September 17, 2019 - 12:45pm-1:10pm

Authors:

Lainscek, D., National Institute of Chemistry
Vida, F., NIC
apela, M., NIC
Jerala, R., National Institute of Chemistry

The CRISPR/Cas system is a highly potent tool which has revolutionized genome engineering and regulation of gene transcription in various cells and organisms. This gene-editing tool consists of a guide RNA (gRNA), which targets the Cas9 endonuclease to the desired genomic site. Cas9 catalyzes the formation of double-strand DNA breaks, which are then repaired by different cell mechanisms. Error-prone Non-homologous end joining occurs, resulting in random indel (insertion-deletion) mutations, which can lead to functional gene inactivation by either frameshift or deletions. Depending on the size (tens of base pairs) of indel mutations, higher rates of â€œknock-outâ€ can be achieved. To achieve greater indel mutations, CRISPR system can be coexpressed in cells with DNA exonucleases, which cause increased recessions of DNA following DNA breaks. We show that joint action of the CRISPR system with different exonucleases significantly increases the percentage of indel mutations at various targeted genes. Of the different exonucleases tested, the *E.coli*-derived exonuclease III (EXOIII) exhibited the best performance in terms of indel formation. To further improve the rate of indel mutations, Cas9 and EXOIII were brought into the proximity via coiled-coil forming heterodimeric peptides (CCExo). This resulted in increased indel formation compared to the classical CRISPR/Cas system as well as more efficient than cointroduction of non-interacting and genetically fused Cas9-EXOIII. We performed a case study for the use of the CRÍSPR-EXO system as a potential anti-cancer therapeutic tool. The Philadelphia chromosome, which occurs in leukemic cancer cells, is the result of characteristic the reciprocal genome translocation t(9:22) and is responsible for higher proliferation of tumorous cells. Using the CCEXO system, we achieved a higher degree of indel mutations at the translocation site, which resulted in greater killing of cancer cells, thus providing a useful potential anti-cancer therapeutic tool.

CRISPR-Powered Transistors for DNA Biosensing

Conference: International Conference on CRISPR Technologies

Year: 2019

Proceeding: 3rd International Conference on CRISPR Technologies

Group: General Submissions

Session: Medical Applications

Time: Tuesday, September 17, 2019 - 9:25am-9:50am

Authors:

Aran, K., Keck Graduate Institute

In recent years, Clustered Regularly Interspaced Short Palindromic Repeats (CRISPR) has garnered a lot of attention for its potential for curative therapies. However, despite its sequence specific targeting capabilities, CRISPR-based biosensors have yet to be developed. We have recently developed a CRISPR powered, electronic biosensor termed, CRISPR-Chip, capable of detecting the binding interactions between nuclease deactivated CRISPR with its target sequence. CRISPR-Chip is a graphene field effect transistor (gFET) where the surface of the graphene is functionalized with nuclease deactivated CRISPR RNA-guided ribonucleoproteins (dRNPs). dRNPs within the CRISPR-Chip construct interact with their target sequence by scanning the whole genomic sample until they find and bind to their target sequence. The selective hybridization of the target DNA to the dRNP complex modulates the grapheneâ€™s conductivity and results in a detectable change in the gFET electrical signal output. CRISPR-Chip was able to detect dRNP binding to two commonly deleted exon target sequences within the human dystrophin gene, without amplification, obtained from clinical samples from patients with confirmed Duchene Muscular dystrophy disease. In addition, preliminary studies have also indicated the potential utilization of CRISPR-Chip for detection of single nucleotide polymorphisms (SNPs) at the target site. CRISPR-chip is the first example of CRISPR-powered electronic transistors that harness the search function of CRISPR/Cas9 and the ultrasensitivity of graphene-based nanoelectronics to enable a complete label-free and amplification-free DNA biosensor.

Enabling Precision Editing with CRISPR Hybrid RNA-DNA Guided Cas9

Conference: International Conference on CRISPR Technologies

Year: 2019

Proceeding: 3rd International Conference on CRISPR Technologies

Group: General Submissions

Session: Medical Applications

Time: Tuesday, September 17, 2019 - 9:50am-10:15am

Authors:

Donohoue, P. D., Caribou Biosciences
Lau, E., Caribou Biosciences
Vidal, B., Caribou Biosciences
Settle, A., Cornell UniversityMemorial Sloan Kettering Cancer Center
Irby, M., Caribou Biosciences
Rotstein, T., Caribou Biosciences
Li, C. H., Caribou Biosciences
Bahn, L., Caribou Biosciences
Toh, M., Caribou Biosciences
Williams, C., Caribou Biosciences
Smith, S., Caribou Biosciences
Gradia, S., Caribou Biosciences
Kohrs, B., Caribou Biosciences
Fuller, C., Caribou Biosciences
van Overbeek, M., Caribou Biosciences
Cameron, P., Caribou Biosciences
Kanner, S. B., Caribou Biosciences

Slorach, E., Caribou Biosciences

The RNA-guided CRISPR-Cas system is a promising technology with applications in a variety of fields including the development of human therapeutics. However, in order to successfully deploy genome editing for clinical applications it is critical to design strategies that maximize on-target activity while minimizing off-target editing. Previously, we developed a comprehensive and robust two-step specificity analysis pipeline that involves (1) identifying off-target sites with our biochemical SITE-Seq® assay and (2) inspecting those sites for editing in cell-based experiments with targeted deep sequencing. We have now applied this pipeline to develop CRISPR hybrid RNA-DNA (chRDNA) guides to attenuate off-target cleavage. Here, we will discuss our chRDNA guide selection workflow which allows us to develop chRDNA guides that exhibit editing efficiencies similar to their crRNA counterparts in human primary T cells and with significantly reduced off-target activity. chRDNA guides enable a high activity, high specificity alternative to the standard system for use in therapeutic applications where the utmost precision is required.

Genome Editing: From Fundamental DNA Repair Mechanisms to Cutting-Edge Applications

Conference: International Conference on CRISPR Technologies

Year: 2019

Proceeding: 3rd International Conference on CRISPR Technologies

Group: General Submissions

Session: Medical Applications

Time: Tuesday, September 17, 2019 - 11:30am-11:55am

Authors:

Kan, Y., eGenesis Inc.

Genome editing: from fundamental DNA repair mechanisms to cutting-edge applications

Yinan Kan[1,2,3], Eric Hendrickson[1,3], George Church[1,2], Luhan Yang[1,2]

[1] eGenesis Inc., 300 Technology Square, Cambridge, MA 02139, USA

[2] Department of Genetics, Harvard Medical School, Boston, MA 02115, USA

[3] Biochemistry, Molecular Biology and Biophysics, University of Minnesota, Minneapolis, MN 55455, USA

Genome editing is the intentional alteration of genetic information in living cells or organisms. Since CRISPR/Cas9 was repurposed for genome editing, the â€œCRISPR Crazeâ€ is quickly bridging the genotype and phenotype worlds and transforming biological and biomedical research. Interestingly, CRISRP/Cas9 does not perform genome editing by itself, but it induces a site-specific genomic lesion and relies on the endogenous DNA repair pathways to introduce desired modifications to the genome.

In this presentation, I will walk through the DNA repair mechanisms of genome editing with a focus on the non-homologous end joining (NHEJ) and homology-directed repair (HDR) pathways. Then I will highlight some of advanced applications of the NHEJ and HDR pathways including multiplex gene knockout and marker-free long-tract gene conversion. Finally, I will introduce eGenesisâ€™ recent efforts in generating the most extensively genome-edited large animals with the aim of providing safe and effective porcine organs for human transplantation.

Improved methods and reagents for CRISPR homology-directed repair

Conference: International Conference on CRISPR Technologies

Year: 2019

Proceeding: 3rd International Conference on CRISPR Technologies

Group: General Submissions

Session: Medical Applications

Time: Tuesday, September 17, 2019 - 11:05am-11:30am

Authors:

Jacobi, A., Integrated DNA Technologies

CRISPR-Cas proteins introduce double-stranded breaks (DSBs) at targeted genomic loci and these are repaired by endogenous cellular pathways such as non-homologous end joining (NHEJ) and homology-directed repair (HDR). Providing a ssDNA template during repair allows researchers to precisely introduce a desired mutation by utilizing the HDR pathway. However, rates of HDR are often low compared to NHEJ-mediated repair. Here, we describe methods to improve the rate of HDR vs NHEJ-mediated repair by careful selection of template characteristics, including homology arm lengths, placement of synonymous SNPs in the HDR template, and the addition of chemical modifications to increase DNA stability against cellular nucleases. In addition, we demonstrate improved HDR rates when using Alt-R HDR Enhancer, a small molecule compound that increases the rate of HDR by inhibiting NHEJ-mediated repair. Increased HDR rates have been achieved in various cell lines, including iPSCs. Furthermore, we implement our findings in the Alt-R CRISPR HDR Design Tool, a novel bioinformatics tool for ssDNA HDR template design. The Alt-R CRISPR HDR Design Tool supports single-stranded designs up to 3 kb, single- and dual-guide designs (i.e., for use with Cas9-Nickases), guide selection suggestions, insertion of synonymous SNPs in coding regions, and more.

Improving CRISPR-Cas Performance with Chemical Modifications in Guide RNAs and Single-Cell DNA Analysis

Conference: International Conference on CRISPR Technologies

Year: 2019

Proceeding: 3rd International Conference on CRISPR Technologies

Group: General Submissions

Session: Medical Applications

Time: Tuesday, September 17, 2019 - 10:40am-11:05am

Authors:

Bruhn, L., Agilent Research Laboratories

As gene editing systems move rapidly from research to therapeutic applications, further improving their performance while better understanding their editing characteristics at the single-cell level becomes fundamentally important. Leveraging our capability to robustly chemically synthesize RNAs of more than 160 nucleotides in length we have developed novel approaches for improving the activity and specificity of CRISPR systems using chemical modifications in guide RNAs. In addition, to mapping positions where chemical modifications can enhance the stability of guide RNAs while maintaining function, we developed a novel approach for enhancing specificity by employing site-specific modifications in the 20-nucleotide DNA recognition sequence in single-guide RNAs (sgRNAs). Our deep sequencing results from human cells transfected with pre-formed Cas9-sgRNA complexes show that a chemical modification (2â €²-O-methyl-3â€²-phosphonoacetate, or â€œMPâ€) incorporated at select sites in the ribose-phosphate backbone of sgRNAs can markedly reduce off-target cleavage activities, while maintaining high on-target performance. Subsequent structural studies showed these positions map onto contacts between Cas9 protein sidechains involved in monitoring the complementarity of the gRNAâ€"target DNA duplex for sufficient helix formation and specificity to allow cleavage.

A more complete understanding of the exact nature of induced mutations in each cell in a population can enable further advances in optimizing the frequencies of on- and off-target gene edits especially when multiple events in each cell are desired. To demonstrate the potential value of single-cell DNA analysis we will present results using the TapestriÂ® platform, a high-throughput, single-cell DNA analysis system which enables the sensitive detection of multiple types of editing events including single nucleotide edits, deletions, insertions, and translocations, and can identify the co-occurrence and zygosity of these changes in different cells in a population of CRISPR edited cells.

Inhibition and Recovery of CRISPR/spCas9 Activity at Closed Chromatin in a Human Cell Line

Conference: International Conference on CRISPR Technologies

Year: 2019

Proceeding: 3rd International Conference on CRISPR Technologies

Group: General Submissions

Session: Medical Applications

Time: Tuesday, September 17, 2019 - 10:15am-10:40am

Authors:

Haynes, K., Emory University
Daer, R., Arizona State University
Hamna, F., Arizona State University
Barrett, C., Arizona State University

To efficiently and reliably edit human DNA in living cells with CRISPR, the Cas9/gRNA complex must overcome barriers imposed by chromatin-mediated compaction. Chromatin is a system of proteins and nucleic acids that organize genomic DNA within the nuclei of eukaryotic cells. Chromatin composed of Polycomb-family proteins mediates epigenetic repression in stem cells and differentiated cells, and is distributed throughout the genome at hundreds of sites that drive cell development pathways. Therefore biomedically important target genes may often be inaccessible to CRISPR. Previously, we demonstrated that closed, silenced chromatin impedes CRISPR editing by making the DNA less accessible to Cas9/gRNA binding. Our investigation of the underlying mechanism showed that disruption of chromatin through siRNA-mediated knock-down of the Polycomb complex enhanced CRISPR editing. Interestingly, target gene overexpression also impeded CRISPR activity. To further investigate the requirements for inducing a CRISPR-accessible state, we compared two approaches: a broad-acting Polycomb-disrupting drug (EZH2 inhibitor UNC1999) and site-specific chromatin-modifying fusion proteins. We explored the mechanism of enhanced editing at artificially opened chromatin in HEK293 cells using deep sequencing of edited DNA and chromatin immunoprecipitation (ChIP) of histone modifications. We demonstrate strong CRISPR enhancement with transient expression of the site-specific, fusion transcriptional activator Gal4P65 and no enhancement with the small molecule chromatin inhibitor UNC1999. In both cases H3K27me3 was depleted from the CRISPR target site, but only Gal4P65 gene-activating enhanced CRISPR efficiency. CRISPR activity was also restored by other Gal4 fusions that did not activate luciferase expression. Our results support the use of site-specific, activator-associated fusion proteins as an effective method to enhance Cas9 editing within Polycomb-repressed chromatin.

New technologies for enhancing the efficiency of CRISPR-based cell engineering at scale

Conference: International Conference on CRISPR Technologies

Year: 2019

Proceeding: 3rd International Conference on CRISPR Technologies

Group: General Submissions

Session: Medical Applications

Time: Tuesday, September 17, 2019 - 9:00am-9:25am

Authors:

Holden, K., Synthego

Synthego has developed automated platforms for engineering immortalized cell lines and stem cells at scale, using the CRISPR-Cas technology. In order to improve the efficiency of genome engineering, new methodologies and technologies are required. Here, we describe innovative technologies and methods for: optimizing and standardizing CRISPR editing methods, temporal control of CRSIRP on and off-targeting, improved tools for predicting editing outcome, and methods for precisely manipulating edited cell populations.

Targeted Transcriptional Modulation with Type I CRISPR-Cas Systems in Human Cells

Conference: International Conference on CRISPR Technologies

Year: 2019

Proceeding: 3rd International Conference on CRISPR Technologies

Group: General Submissions

Session: Medical Applications

Time: Tuesday, September 17, 2019 - 12:20pm-12:45pm

Authors:

Pickar-Oliver, A., Duke University
Black, J. B., Duke University
Lewis, M. M., Duke University
Mutchnick, K. J., Duke University
Klann, T. S., Duke University
Gilcrest, K. A., Duke University
Sitton, M. J., Duke University
Nelson, C. E., Duke University
Barrera, A., Duke University
Bartelt, L. C., Duke University
Reddy, T. E., Duke University
Beisel, C. L., Helmholtz Institute for RNA-based Infection Research
Barrangou, R., North Carolina State University
Gersbach, C. A., Duke University

The CRISPR revolution has highlighted bacterial adaptive immune systems as a rich and largely unexplored frontier for discovery of new genome engineering technologies. In particular, Cas9, a class 2 CRISPR-Cas systems, has been widely applied for targeting DNA sequences in eukaryotic genomes. However, the great diversity of CRISPR systems existing throughout the natural world has not yet been explored for potential genome engineering tools in human cells. Here, we report DNA-targeting and transcriptional control with class 1 CRISPR-Cas systems. Specifically, we repurposed the Cascade effector complex from type I variants of class 1 CRISPR-Cas systems, the most prevalent CRISPR loci in nature. Cascade targets double-stranded DNA via a multi-component protein complex and a single CRISPR RNA (crRNA). We first designed codon-optimized mammalian expression cassettes for each Cascade subunit from the *E. coli* strain K12 (termed EcoCascade). Following transfection of human cells, we validated expression of each subunit, proper nuclear localization, and Cascade complex formation. We demonstrated programmable crRNA-mediated targeting of specific loci in the human genome via ChIP-seq analysis. By tethering transactivation domains to Cascade, we modulated the expression of multiple targeted chromosomal genes in human cells, including >3,000-fold *IL1RN*activation comparable to activation with dCas9-based effectors. Transcriptome-wide analysis by RNA-seq showed highly specific crRNA-dependent activation of the targeted gene. To generalize this approach to other Cascade complexes, we applied our understanding of the model type I-E EcoCascade to the type I-B CRISPR-Cas system of *L. monocytogenes* Finland_1998 (termed LmoCascade). Following tethering of transactivation domains to LmoCascade, we showed endogenous gene activation in human cells with almost all tested crRNAs. Additionally, we tethered the KRAB effector to LmoCascade and demonstrated robust targeted transcriptional repression at endogenous loci. This study expands the toolbox for engineering mammalian genomes and establishes Cascade as a novel CRISPR-based technology for targeted mammalian gene regulation.

Targeting monogenetic liver diseases by CRISPR/Cas9

Conference: International Conference on CRISPR Technologies

Year: 2019

Proceeding: 3rd International Conference on CRISPR Technologies

Group: General Submissions

Session: Medical Applications

Time: Tuesday, September 17, 2019 - 11:55am-12:20pm

Authors:

Schwank, G., ETH Zurich

Class 2 type V CRISPR-Cas systems are characterized by a single effector nuclease which requires guide RNA for activity, making them attractive genome-editing tools. A distinct group of type V effector variants, termed Cms1, has recently been identified in metagenomic datasets. Their size and domain architecture set them apart from other known type V nucleases, suggesting likely functional differences. Many of the more than 40 predicted putative Cms1 nucleases exhibit editing activity in plants. However, catalytic properties and biotechnological potential of these effectors remain largely unexplored. In this study we characterized the SuCms1 nuclease identified in a metagenomic sequence of an uncultured chemolithoautotrophic epsilonproteobacterium, Sulfuricurvum spp., from the Frasassi Cave, Italy. We carried out our assessment of the fundamental properties of SuCms1 in Escherichia coli BL21-AI heterologous host. SuCms1 exhibited two nuclease activities, mediating CRISPR-RNA biogenesis and catalyzing RNA-dependent interference by cleaving dsDNA targets. This cleavage was dependent on the three RuvC domains, similar to Cas12a. However, unlike any know type V nuclease, SuCms1 exhibited a broad PAM preference, recognizing approximately half of the nucleotide combinations in a 5N library. Furthermore, the role of the -5 position, which is not part of the Cas12a consensus PAM, was context dependent in the case of SuCms1. Finally, we showed that known type V anti-CRISPR (AcrVA) proteins did not inhibit SuCms1. Particularly notable is the lack of inhibition by AcrVA5 which inactivates Cas12a by acetylating a conserved PAM recognition lysine residue also found in SuCms1. These data suggest that the PAM recognition mechanism of SuCms1 is entirely different from that of Cas12a, potentially relying on a combination of sequence identities and associated structural DNA features. This work helps establish SuCms1 as a unique variant among type V CRISPR-nucleases and informs efforts of adopting Cms1 effectors for biotechnological applications.

Genome Editing for Consumer Traits in Fresh Produce

Conference: International Conference on CRISPR Technologies

Year: 2019

Proceeding: 3rd International Conference on CRISPR Technologies

Group: General Submissions

Session: Agricultural Applications

Time: Tuesday, September 17, 2019 - 3:55pm-4:20pm

Authors:

Fraebel, S., Pairwise Plants

Nutritional deficit is the leading mortality risk factor globally. Consumption of fruits and vegetables â€“ the most effective solution to malnutrition â€“ can be limited by low production or high cost in some countries and by poor quality characteristics or inconvenient preparation in others. Americans, for example, consume only half the recommended daily intake of fruits and vegetables despite widespread availability. Pairwise seeks to address these challenges regionally and globally by making healthy food more convenient, affordable, and sustainable. To fulfill this mission, we have assembled a world class portfolio of CRISPR technologies and a state-of-the-art R&D capability focused on delivering differentiated products to the fresh produce aisle. We have established base editing and cutting tools with industry-leading performance profiles in plants, including editing efficiencies as high as 85%. We are applying these tools for novel consumer traits in cane berries.

How to Turn Negative into Positive: Application of CRISPR Endoribonuclease Csy4 As a Negative Operator to Detect Negative Signals in Planta

CRISPR TECHNOLOGIES CONFERENCE

Conference: International Conference on CRISPR Technologies

Year: 2019

Proceeding: 3rd International Conference on CRISPR Technologies

Group: General Submissions

Session: Agricultural Applications

Time: Tuesday, September 17, 2019 - 3:30pm-3:55pm

Authors:

Schreiber, T., Leibniz Institute of Plant Biochemistry
Prange, A., Leibniz Institute of Plant Biochemistry
Tissier, A., Leibniz Institute of Plant Biochemistry

The CRISPR Class 1 endoribonuclease Csy4 from *Pseudomonas aeruginosa* binds and cleaves the repetitive sequence of the pre-crRNA. The Csy4 target sequence (C4T) can be integrated into reporter genes (like GUS or Venus) making them Csy4-sensitive on the RNA-level. *In planta* studies revealed that Csy4 can completely block reporter gene expression. This served as basis for the investigation of RNA-degrading and DNA-editing tools targeting Csy4. This system allows to convert a gene-repressing component into a positive signal. Thus, the optimization of these repressing components can be carried out by monitoring the extent of de-repression of the reporter. Here, properties and applications of the Csy4-based reporter system will be presented.

Population genome editing for the control of the human malaria vector Anopheles gambiae

Conference: International Conference on CRISPR Technologies

Year: 2019

Proceeding: 3rd International Conference on CRISPR Technologies

Group: General Submissions

Session: Societal Impacts

Time: Wednesday, September 18, 2019 - 9:00am-9:25am

Authors:

Crisanti, A., Imperial College

Mosquitoes species of the genus Aedes and Anopheles are responsible for transmitting severe and life threatening diseases including a number of viral encephalitis, Dengue yellow fever, Malaria and more recently Zika. A few Anopheles species are responsible for causing 200 million cases of malaria every year and the death of half a million children under the age of five in less developed regions of Africa. During the last twenty years a worldwide concerted effort based on the use of bed nets, insecticides and drug treatment has halved malaria morbidity and mortality. The implementation of these control measures necessitates about 10 billion per year mostly in the form of donations thus questioning the long-term sustainability of this approach and its suitability for eradicating the disease in the next 30-40 years. The vectorial capacity of a mosquito species to transmit malaria depends on genetically determined traits such as feeding behaviour, longevity, density and ability to support parasite development. Editing of the corresponding genes is anticipated to impair mosquito ability to transmit malaria. The recent development of CRISPR/CAS9 based gene drive technology has unlocked the possibility to selectively edit a mosquito population. Genetic modifications designed to either impair female fertility or interfere with mosquito ability to transmit the malaria parasite have been spread from few laboratory individual to large caged mosquito populations. These laboratory experiments have also supported mathematical modelling predicting how gene technology has the potential to eradicate malaria transmission in a span of few years from vast regions of Africa. Technical challenges in the development of a gene drive technology suitable for release include the development nuclease-resistant functional gene variant that would block the spreading of the drive as well as off target activity of the CAS9 nuclease that may generate undesirable mutations at other loci.

A Minimal CRISPR-Cas3 System for Rapid Large-Scale Genome Manipulation

Conference: International Conference on CRISPR Technologies

Year: 2019

Proceeding: 3rd International Conference on CRISPR Technologies

Group: General Submissions

Session: Bacterial Applications

Time: Wednesday, September 18, 2019 - 11:00am-11:25am

Authors:

Leon, L., UCSF
Bondy-Denomy, J., UCSF
Csorgo, B., UCSF
Chau, I., UCB
Lewis, J., UCB

CRISPR-Cas systems are a diverse branch of prokaryotic immune systems that have been co-opted for prokaryotic and eukaryotic gene editing. Here, we present CRISPR-Cas3 based editing, an easily-programmable genome engineering technique that allows for the rapid generation of large genomic deletions. Cas3, the signature enzyme of CRISPR-Cas Type I systems, possesses coupled nuclease and helicase activity, allowing for processive degradation of a targeted region. While the leading CRISPR-Cas tools (e.g. Cas9 and Cas12a) allow for genetic edits like small insertions and deletions, these enzymes are limited when the desired manipulation scales in size (>10 kb). With this in mind, we optimized a CRISPR-Cas3 system to generate both random and pre-designed large deletions (up to 250 kb) in *Pseudomonas aeruginosa*. Iterative transformation of one strain with ten distinct guides yielded ten deletions resulting in a reduction of 10% of the genome in thirty days. Additionally, using one guide to target loci that exist in multiple copies within a single chromosome allowed us to simultaneously delete more than one portion of the genome. To test the portability of CRISPR-Cas3, we induced large deletions in both *Pseudomonas syringae* and*Escherichia coli* using an â€œall-in-oneâ€ vector expressing *cas* genes and a guide. Deletion efficiencies in these orthogonal systems remained high. Thus, a tool like CRISPR-Cas3 will facilitate rapid strain manipulation for synthetic biological and metabolic engineering purposes as well as the analysis of regions of unknown function in diverse organisms for basic biological research.

A Simple and Fast Method for Transcriptional Inactivation in Synechocystis PCC6803 Using CRISPR-Interference from a Single Plasmid

Conference: International Conference on CRISPR Technologies

Year: 2019

Proceeding: 3rd International Conference on CRISPR Technologies

Group: General Submissions

Session: Bacterial Applications

Time: Wednesday, September 18, 2019 - 11:25am-11:50am

Authors:

Kirtania, P., Institute of Plant Biology, Biological Research Centre, Hungarian Academy of Sciences
Hadi, B., Institute of Plant Biology, Biological Research Centre, Hungarian Academy of Sciences
Mallick, I., Institute of Plant Biology, Biological Research Centre, Hungarian Academy of Sciences
Vass, I. Z., Institute of Plant Biology, Biological Research Centre, Hungarian Academy of Sciences
Fehar, T., Institute of Biochemistry, Biological Research Centre, Hungarian Academy of Sciences
Vass, I., Institute of Plant Biology, Biological Research Centre, Hungarian Academy of Sciences
B. Kas, P., Institute of Plant Biology, Biological Research Centre, Hungarian Academy of Sciences

Synechocystis PCC6803 is an important organism for its biotechnological applications and in studying photosynthesis. However, due to the dynamic nature of its chromosome number, creating a desired mutant is a tedious process. Recently, CRISPR-interference using deadCas9 was applied to create mutants in *Synechocystis* PCC6803 to speed up the process, however, as this method includes genomic integration of deadCas9-cassette, it is not lesion free [1]. Here, we show a simple, fast and lesion free method to apply CRISPR-interference for transcriptional inactivation in *Synechocystis* PCC6803 using the single shuttle-plasmid (pCRPB1010). The technique is easier to execute for gene silencing in*Synechocystis* PCC6803 as it circumvents the genome integration and segregation steps and thereby significantly shortens the construction of the mutant strains.

To validate the technique, we targeted well characterized genes involved in photosynthesis to get a clear phenotype. First, we targeted the non-template DNA strand of *psbO* using a single spacer. 70% reduction in PsbO expression and about 50% reduction of oxygen evolution were observed compared to the wild type. Further, targeting the common span of the template DNA strand of *psbA2 and psbA3* genes using a single spacer resulted in a complete abolition of D1 protein expression, and complete loss of both oxygen evolution and photo-autotrophic growth was observed. When the spacers were combined could achieve significant extent of silencing of multiple genes stated above using one common plasmid construct.

This is a first report of a method developed for applying CRISPR-interference in a completely lesion free single-plasmid based episomal manner in *Synechocystis* PCC603.

1. Yao, L., et al., *Multiple Gene Repression in Cyanobacteria Using CRISPRi.* ACS Synth Biol, 2016. **5**(3): p. 207-12.

Enhancing Bacterial Crispri Specificity Using Active ThermoCas9 and an Anticrispr Protein

Conference: International Conference on CRISPR Technologies

Year: 2019

Proceeding: 3rd International Conference on CRISPR Technologies

Group: General Submissions

Session: Bacterial Applications

Time: Wednesday, September 18, 2019 - 11:50am-12:15pm

Authors:

Mougiakos, I., Wageningen UR

CRISPR-Cas9 technologies have enabled unprecedented genome editing and transcriptional regulation efficiency in prokaryotes, accelerating the exploration and exploitation of their metabolism. However the target-DNA binding specificity of the extensively employed *Streptococcus pyogenes* (Sp)Cas9, as well as of currently reported engineered variants, is low. This fact compromises the accuracy of CRISPRi-based metabolic engineering studies in bacteria. Here we report the *in vitro* (at 37°C) and*in vivo* (in *Escherichia coli*) characterization of the thermo-tolerant Cas9 orthologue from *Geobacillus thermodenitrificans* T12 (ThermoCas9) that shows enhanced DNA binding and targeting specificity in comparison to previously reported data for other Cas9 orthologs as well as improved DNA binding specificity compared to its catalytically inactive variant, Thermo-dCas9. To further exploit the ThermoCas9 binding specificity for CRISPRi purposes, we harness the function of a small anti-CRISPR protein from *Neisseria meningitidis* ($\text{AcrIIC1}_{\text{Nme}}$). We show that $\text{AcrIIC1}_{\text{Nme}}$ traps ThermoCas9 *in vivo* in a DNA-bound, catalytically inactive state, robustly inhibiting targeting and resulting in transcriptional silencing with comparable on-target efficiency to Thermo-dCas9.

Silencing Genes with dCas9: A Story of Fine-Tuned Control, High-Throughput Screens and Surprising Toxicity

Conference: International Conference on CRISPR Technologies

Year: 2019

Proceeding: 3rd International Conference on CRISPR Technologies

Group: General Submissions

Session: Bacterial Applications

Time: Wednesday, September 18, 2019 - 10:10am-10:35am

Authors:

Bikard, D., Institut Pasteur

The catalytically dead variant of Cas9 known as dCas9 can be guided by small RNAs to block transcription of target genes, in a strategy also known as CRISPRi. The level of complementarity between the guide RNA and the target controls the rate at which RNA polymerase 'kicks out' dCas9 from the target and completes transcription. We used this mechanism to precisely and robustly reduce gene expression by defined relative amounts. Silencing genes with dCas9 can also be performed in high-throughput screens to decipher gene functions and genetic interactions. We recently used a genome-wide library of guide RNAs to direct dCas9 to block gene transcriptions in *E. coli*. The results can be used to investigate gene essentiality and provide insights into the effect of dCas9 binding position and orientation, as well as its polar effects on gene expression. Surprisingly, we revealed that guide RNAs sharing specific 5 nucleotides seed sequences can produce strong fitness defects and even kill *E. coli* regardless of the other 15nt of guide sequence. This effect which occurs at high dCas9 concentrations can be alleviated by tuning the expression of dCas9 while maintaining strong on-target repression. Altogether our work demonstrates the usefulness of CRISPRi in bacteria while providing important design rules.

The Role of NHEJ Double-Strand Break Repair on CRISPR-Cas9 Targeting in Pseudomonas Putida KT2440

Conference: International Conference on CRISPR Technologies

Year: 2019

Proceeding: 3rd International Conference on CRISPR Technologies

Group: General Submissions

Session: Bacterial Applications

Time: Wednesday, September 18, 2019 - 10:35am-11:00am

Authors:

Chacan Machado, L., Skolkovo Institute of Science and Technology
Savitskaya, E., Skolkovo Institute of Science and Technology
Severinov, K., Skolkovo Institute of Science and Technology

A novel three-plasmid system of targeted genomic engineering in*Pseudomonas putida* KT2440 has been developed in our laboratory. It includes ssDNA recombineering followed by Cas9 counterselection[1] . So far, point mutations and deletions have been performed by our method, however the efficiency of counterselection has proven to be low in the majority of targeted loci. In contrast to most prokaryotes, *P. putida*possesses NHEJ associated genes, encoding for Ku and LigD proteins, whose activity has been associated with stationary-phase mutagenesis in starving *P. putida*[2] .

In this work we (1) investigated whether the low counterselection efficiency is caused by a non-templated error prone repair of the double-stranded break (DSB) performed by Cas9-sgRNA efector and (2) evaluated the posible role of Ku and LigD proteins in the process.

We found that, unlike many other bacteria, *P. putida* can overcome the Cas9-associated toxicity by repairing the DSB in error prone way. Even more, knocking/overexpressing *ku* or *ligD* genes did not affect the counterselection efficiency but did change the mutation spectrum in the target region, suggesting that, besides the canonical NHEJ, other alternative pathways may be involved in the repair event.

The understanding of how DSBs are repaired via NHEJ in *P. putida* will contribute to the optimization of editing strategies based on CRISPR/Cas9 counterselection.

[1] Quintana Cordero J. ssDNA recombineering followed by Cas9/sgRNA counterselection as a new tool for *Pseudomonas putida* genome editing. Masterâ€™s Thesis, Skolkovo Institute of Science and Technology, 2018.

[2] Paris U., Mikkel K., Tavita K., Saumaa S., Teras R., Kivisaar M. (2015) NHEJ enzymes LigD and Ku participate in stationary-phase mutagenesis in Pseudomonas putida. DNA Repair 31, 11-18.

CRISPR and Zika Eradication a Multidisciplinary and Anticipatory Ethical Analysis

Conference: International Conference on CRISPR Technologies

Year: 2019

Proceeding: 3rd International Conference on CRISPR Technologies

Group: Poster Session

Session: Poster Session

Authors:

Wilson, R., Towson University/University of Baltiore

Microcephaly and other birth defects could be the tip of the iceberg," Dr. Sonja Rasmussen of the U.S. Centers for Disease Control and Prevention said at the annual meeting of the Pediatric Academic Societies. "The true burden of congenital disease with Zika virus is probably underestimated," said Dr. Marco Safadi of the Santa Casa Medical School in Sao Paulo, Brazil, who's been treating and studying cases. Zika can be caused by mosquitos and by sexual transmission.

Zika may stay around for years, citing a study just published that showed monkeys have been infected with Zika in Brazil. That means they can act as a reservoir for the virus, as they do with yellow fever.

Even if a vaccine is developed and people get vaccinated, or if enough of the population gets infected to confer widespread immunity, the monkeys will be there for the mosquitoes to bite and carry the infection back into people in the future. This will make Zika a problem for years to come.

The question is how to control this disease vector.

Ethics involves the intentions, actions and the outcomes of actions as performed by actors. Anticipatory ethics is concerned with identifying potential problems with technology while it is in the early stages of development. In this paper we will extend the focus of anticipatory ethics to include the use of technology CRISPR genetic engineering technology to the project of genetically modifying mosquitos . The analysis will explore how CRISPR technology is now being used to modify mosquito populations in order to address human health care problems and examine how the introduction of genetically modified mosquitos in Brazil has led to an end of the Zika health emergency.

CRISPR/Cas9-Mediated Target Validation of the Anti-Cancer Drug Pladienolide B

CRISPR TECHNOLOGIES CONFERENCE

Conference: International Conference on CRISPR Technologies

Year: 2019

Proceeding: 3rd International Conference on CRISPR Technologies

Group: Poster Session

Session: Poster Session

Authors:

Aouida, M., Hamad Bin Khalifa University

CRISPR/Cas9 system confers molecular immunity in archeal and bacterial species against invading foreign nucleic acids. CRISPR/Cas9 system is used for genome engineering applications across diverse eukaryotic species. In this study, we demonstrate the utility of the CRISPR/Cas9 genome engineering system for drug target validation in human cells. Pladienolide B is a natural macrolide with antitumor activities mediated through the inhibition of pre-mRNA splicing. To validate the spliceosomal target of Pladienolide B, we employed the CRSIPR/Cas9 system to introduce targeted mutations in the subunits of the SF3B complex in the HEK293T cells. Our data reveal that targeted mutagenesis of the SF3b1 subunit exhibited higher levels of resistance to Pladienolide B. Therefore, our data validate the spliceosomal target of Pladienolide B and provide a proof of concept on using the CRISPR/Cas9 system for drug target identification and validation.

Head-to-Tail Multiplications of DNA Templates at Sites of Integration Mask Desired CRISPR/Cas9 - Mediated Genome Editing Events

Conference: International Conference on CRISPR Technologies

Year: 2019

Proceeding: 3rd International Conference on CRISPR Technologies

Group: Poster Session

Session: Poster Session

Authors:

Skryabin, B. V., Core Facility Transgenic animal and genetic engineering Models (TRAM)
Rozhdestvensky, T. S., Core Facility Transgenic animal and genetic engineering Models (TRAM)
Roth, J., Institute of Immunology
Meuth, S. G., Clinic of Neurology with Institute of Translational Neurology
Pavenstadt, H., Internal Medicine D
Sherwood, J., Institute of Experimental Musculoskeletal Medicine (IMM)
Pap, T., Institute of Experimental Musculoskeletal Medicine (IMM)
Wedlich-Saldner, R., Institute of Cell Dynamics and Imaging
Sunderkatter, C., Department of Dermatology and Venereology
Schwartz, Y. B., Department of Molecular Biology
Brosius, J., Institute of Experimental Pathology (ZMBE)

CRISPR/Cas9 mediated genome editing has rapidly evolved into a key technology for precise gene editing in a wide range of model organisms, including mouse and human. Nowadays, CRISPR/Cas9 is widely applied for generation of animal models to investigate different human diseases. Despite the advantages of CRISPR/Cas9 based genome editing, a number of potential problems such as genome rearrangements and off target effects still impede the CRISPR/Cas9 technology for use in biomedical research and further efforts are necessary to overcome these hurdles.

Our study examines problems that affect direct knock-in genome targeting. During the generation of six different conditional knock-out mouse models, we discovered that frequently (sometimes solely) homology-directed repair and/or non-homologous end-joining mechanisms caused multiple unwanted head-to-tail insertions of donor DNA templates. Disturbingly, conventionally applied PCR analysis - in most cases - failed to identify such multiple integration events, which led to a high rate of falsely claimed precisely edited alleles. We caution that comprehensive analyses of modified alleles are essential, and offer practical solutions to correctly identify precisely edited chromosomes.

Our findings are important to unlock the full potential of the CRISPR/Cas9-mediated genome editing protocols for the generation of custom designed gene variants for biomedical research and gene therapy.

Prediction of Pesticides Residues Quantities in a Proposed Food and Suggestion of Suitable Utilization for Pesticides in a Long-Term Usage

Conference: International Conference on CRISPR Technologies

Year: 2019

Proceeding: 3rd International Conference on CRISPR Technologies

Group: Poster Session

Session: Poster Session

Authors:

Noureldien, M., University of Khartoum

Pesticides are defined as a substance or a mixture of substances used for preventing, destroying, repelling, or mitigation of any pest. They are Usually classified by the intended use, level of toxicity however this article is concerned with the classification by the type of the pest they control. It has been proved that from the millions of tons of the utilized pesticide only 1% reach the targeted pest. The rest appear as residues in food, soil, water in addition to air and often linked these residues with long-term side effects including neurological problems, bird defects, neurodevelopmental disorder, environmental impact on animals, cancer and weather. Despite the fact that most of the pesticides have a relative short effect on the targeted region and the targeted pests itself, only the chlorinated hydrocarbons (e.g. chlordane, Aldrin, DDT, HCH), in addition to limited organophosphorus pesticides (e.g. carbophenothion) have a known long residual action. On the other hand, it is difficult to predict whether a proposed new pesticide can or canâ€™t cause these impacts since they are long term side effects. This article propose a numerical method termed as *(Numerical Oriented Usage Regulator)* to predict whether a proposed pesticide is significantly likely to make hazard residues in selected sort of food and suggest the regulations of the utilization of the proposed Pesticide. The US PDPâ€™s (Pesticides

Data Program) data from 1992 till 2017 were used as a data source to implement and test the proposed method. The objective of this article to assist the scientists to predict the possibility of a proposed pesticide to make above-limited residues in food which is a long term side effect and hard to predict, by using a relatively short term scale.

Programmed DNA Destruction By Miniature CRISPR-Cas14 Enzymes

Conference: International Conference on CRISPR Technologies

Year: 2019

Proceeding: 3rd International Conference on CRISPR Technologies

Group: Poster Session

Session: Poster Session

Authors:

Harrington, L., UC Berkeley
Burstein, D., Tel Aviv University
Chen, J. S., University of California Berkeley
Paez, D., Universidad de los Andes
Cofsky, J., UC Berkeley
Banfield, J. F., University of California
Doudna, J. A., University of California Berkeley

CRISPR-Cas systems provide microbes with adaptive immunity to infectious nucleic acids and are widely employed as genome editing tools. These tools use RNA-guided Cas proteins whose large size (950 to 1400 amino acids) has been considered essential to their specific DNA- or RNA-targeting activities. Here we present a set of CRISPR-Cas systems from uncultivated archaea that contain Cas14, a family of exceptionally compact RNA- guided nucleases (400 to 700 amino acids). Despite their small size, Cas14 proteins are capable of targeted single-stranded DNA (ssDNA) cleavage without restrictive sequence requirements. Moreover, target recognition by Cas14 triggers nonspecific cutting of ssDNA molecules, an activity that enables high-fidelity single-nucleotide polymorphism genotyping (Cas14-DETECTR). Metagenomic data show that multiple CRISPR-Cas14 systems evolved independently and suggest a potential evolutionary origin of single- effector CRISPR-based adaptive immunity

A Cas12a-Mediated Crispri Screen for Small Noncoding RNAs in the Human Commensal Bacteroides Thetaiotaomicron

Conference: International Conference on CRISPR Technologies

Year: 2019

Proceeding: 3rd International Conference on CRISPR Technologies

Group: Poster Session

Session: Registered Posters

Authors:

Prezza, G., Helmholtz Institute for RNA-based Infection Research
Ryan, D., Helmholtz Institute for RNA-based Infection Research
Liao, C., Helmholtz Institute for RNA-based Infection Research
Beisel, C., Helmholtz Institute for RNA-based Infection Research
Westermann, A. J., Helmholtz Institute for RNA-based Infection Research

Gram-negative, obligate anaerobic *Bacteroides thetaiotaomicron* a dominant member of the intestinal bacterial consortium in healthy human beings. In the past, transposon mutant libraries of *B. thetaiotaomicron* were screened to identify genetic determinants required for colonization of germ-free (Goodman et al, CH&M, 2009) and gnotobiotic (Wu et al, Science, 2015) mice. This approach, whose underlying principle is that the insertion of a transposon disrupts the function of the targeted gene, has however been largely restricted the screening of protein-coding genes and operons.

Small noncoding RNAs (sRNAs) are widely used within bacteria to adapt gene expression in response to myriads of external and intrinsic cues. We recently identified hundreds of sRNA candidates in the *B. thetaiotaomicron* genome, whose functions are currently not known. However, due to their short length (typically 50-500 nt), sRNAs are less likely to be hit by random mutagenesis and, consequently, are poorly reflected in existing Tn-seq (transposon sequencing) datasets.

For this reason, in this work we seek to employ CRISPR interference (CRISPRi) to identify and characterize *B. thetaiotaomicron* sRNAs essential for growth under defined conditions. We *in-silico* screened several protospacer adjacent motifs (PAMs) to identify the one with the highest occurrence among the sRNA pool. The analysis resulted in the AT-rich 5â€™-TTV PAM recognized by Cas12a (Cpf1), in line with the low GC content of the *B. thetaiotaomicron* genome. We also employed a cell-free transcription and translation system (TXTL) to screen potential nucleases to express with our library. Finally, by using a luciferase reporter, we assessed how the position and strandness of the targeted region inside the reporter gene affect the extent of transcriptional repression by Cas12a.

The CRISPRi approach described in this study will allow us to pinpoint functionally important sRNAs in *B. thetaiotaomicron* grown under defined conditions, including the ones that mimic the bacteriumâ€™s *in-vivo* niche.

A Novel One-Step Method for Targeted Multiplication of DNA Fragments in Escherichia coli chromosome

Conference: International Conference on CRISPR Technologies

Year: 2019

Proceeding: 3rd International Conference on CRISPR Technologies

Group: Poster Session

Session: Registered Posters

Authors:

Igonina, O., Closed Joint-Stock Company Ajinomoto-Genetika Research Institute
Hook, C., Closed Joint-Stock Company Ajinomoto-Genetika Research Institute
Stoynova, N., Closed Joint-Stock Company Ajinomoto-Genetika Research Institute
Samsonov, V., Closed Joint-Stock Company Ajinomoto-Genetika Research Institute
Krylov, A., Closed Joint-Stock Company Ajinomoto-Genetika Research Institute

E. coli is a well-characterized bacterium widely used as a workhorse in industrial biotechnology, synthetic biology, and metabolic engineering. Therefore, a wide variety of tools for *E. coli* chromosome editing was developed. However, the development of â€œeasy-to-useâ€ methods of insertion of the large chromosomal fragments into specific loci in multiple copies is still of current importance.

The CRISPR-Cas9 based methods are straightforward, efficient and reliable for making a single nucleotide mutations or deletions (Jiang, 2013). However, in case of targeted insertion of DNA fragments, CRISPR-Cas9-assisted tools require plasmid which will be a donor of genes for insertion (Jiang, 2015). The construction of a donor plasmid may be laborious and time-consuming task, especially in case of a long-length DNA fragment and/or a synthetic construct which expression as part of a plasmid is strongly undesirable.

We designed a novel technique for *E. coli* chromosome editing, which allows a targeted stable multiplication of a chosen long chromosome fragment in one-step experiment: the chromosome region of interest is *in vivo* subcloned via Î»Red-recombination into the short synthetic non-replicable DNA fragment containing the excisable marker and Ï†80-*attP*site. The resulting circle DNA molecule is immediately inserted into alternative chromosomal locus due to Ï†80-Integrase activity; the possibility for integration into several pre-existed Ï†80-*attB* sites simultaneously was demonstrated. In the proposed method, multiple Ï•80-*attB* recombination sites can be introduced into a recipient strain with CRISPR-Cas9 methodology.

Jiang, Wenyan, et al. "RNA-guided editing of bacterial genomes using CRISPR-Cas systems." *Nature biotechnology* 31.3 (2013): 233.

Jiang, Yu, et al. "Multigene editing in the Escherichia coli genome via the CRISPR-Cas9 system." *Appl. Environ. Microbiol.* 81.7 (2015): 2506-2514.

Adaptation By T. Thermophilus Type III CRISPR-Cas System in the Presence of Different Phages and Appearance of Viral Escapes

Conference: International Conference on CRISPR Technologies

Year: 2019

Proceeding: 3rd International Conference on CRISPR Technologies

Group: Poster Session

Session: Registered Posters

Authors:

Artamonova, D., Skolkovo Institute of Science and Technology
Karneyeva, K., Skolkovo Institute of Science and Technology
Severinov, K., Skolkovo Institute of Science and Technology
Yasinskaya, A., Skolkovo Institute of Science and Technology
Milenkin, A., Moscow Institute of Physics and Technology

In our work, we study adaptation by Type III CRISPR-Cas system usingThermus thermophilus Hb27 strain as a model organism. This hyperthermophilic bacterium encodes CRISPRâ€"Cas systems of IIIâ€"A, IIIâ€"B, Iâ€"B and Iâ€"C subtypes. It has ten CRISPR arrays of three different types. Six of these arrays belong to type III (they can be shared between both IIIâ€"A and IIIâ€"B subtypes).

We found that infection of T. thermophilus Hb27 cells with a siphovirus phiFa resulted in a very biased adaptation. The majority of acquired spacers originated from the region of the phage genome that encodes its early genes. In contrast, adaptation in conditions of infecting T. thermophilus Hb27 with a tectivirus phiKo led to a uniform distribution of the acquired spacers over the entire genome. In the case of phiKo, spacers against each of the temporal classes of genes enabled an efficient interference. In the case of phiFa, we have not detected any middle or late gene targeting spacers that would protect cells, suggesting that there may be a phage-encoded anti-CRISPR mechanism. After re-infecting the resistant strains bearing spacers against phiFa, we detected some escape phages with lesions in protospacers, revealing the requirements for productive spacer-protospacer interactions for the Type III CRISPR-Cas system.

Analysis of Spacers Precursors in Cells Undergoing Naive Adaptation

Conference: International Conference on CRISPR Technologies

Year: 2019

Proceeding: 3rd International Conference on CRISPR Technologies

Group: Poster Session

Session: Registered Posters

Authors:

M. Guzman, N., University of Alicante
Savitskaya, E., Skolkovo Institute of Science and Technology
Musharova, O., Institute of Molecular Genetics, Russian Academy of Sciences
Severinov, K., Skolkovo Institute of Science and Technology
J.M.Mojica, F., University of Alicante

During primed CRISPR adaptation, spacers precursors of proper length are generated by the Cas3 nuclease-helicase, a component of CRISPR interference machinery that degrades targets recognized by CRISPR effector complex [1]. Naïve adaptation occurs in cells that do not contain spacers matching a target and is critical for initial acquisition of immunity [2,3]. Several studies in *Escherichia coli* suggest that there is a functional link between the RecBCD recombination complex and Chi DNA sequences it recognizes and spacer acquisition during naïve adaptation [4,5]. DNA fragments generated by RecBCD are thought to be incorporated, at low frequency, into CRISPR array by the Cas1-Cas2 adaptation complex as spacers. The mechanism by which these fragments are selected and processed for incorporation into CRIPR array is not fully understood. To address this issue, we have performed ChIP-Sequencing of *Escherichia coli* cells undergoing naïve adaptation to reveal DNA fragments associated with the Cas1-Cas2 complex. Our preliminary data suggest that Cas1-Cas2 associated precursor fragments in cells undergoing naÃ¯ve adaptation are much longer than spacer-sized fragments observed in cells undergoing primed adaptation. A direct correlation between abundance of precursor fragments and number of spacers acquired from them is observed. Further analysis using *recBCD* mutants will reveal the mechanism by which the precursors of spacers acquired during naÃ¯ve adaptation are generated.

[1] Musharova, O. Spacer-length DNA intermediates are associated with Cas1 in cells undergoing primed CRISPR adaptation. NAR 2017, 45(6), 3297â€"3307.

[2] Sternberg SH. Adaptation in CRISPR-Cas Systems. Mol Cell. 2016 Mar 17;61(6).

[3] Datsenko KA. Molecular memory of prior infections activates the CRISPR/Cas adaptive bacterial immunity system. Nat Commun 2012; 3:945.

[4] Levy, A. CRISPR adaptation biases explain preference for acquisition of foreign DNA. Nature 2015, 520(7548), 505â€"510.

[5] IvanÄiÄ‡-Bace, I. Different genome stability proteins underpin primed and NaÃ¯ve adaptation in E. Coli CRISPR-Cas immunity. NAR 2015 43(22).

Binding Does Not Equal Function: Linking Glucocorticoid Receptor Binding to the Regulation of Genes in the Endogenous Genomic Context

CRISPR TECHNOLOGIES CONFERENCE

Conference: International Conference on CRISPR Technologies

Year: 2019

Proceeding: 3rd International Conference on CRISPR Technologies

Group: Poster Session

Session: Registered Posters

Authors:

Meijsing, S., Max Planck Unit for the Science of pathogens
Thormann, V., Max Planck Institute for Molecular Genetics
Vingron, M., Max Planck Institute for Molecular Genetics
Charpentier, E., Max Planck Unit for the Science of Pathogens
Schapflin, R., Max Planck Institute for Molecuar Genetics

The glucocorticoid receptor (GR) binds to tens of thousands of genomic binding sites, yet seems to regulate a much smaller number of genes. In a simple scenario, binding of GR would influence the expression of associated genes whereas in the absence of binding, nearby genes would not be affected. In practice however, the link between GR binding and gene regulation is fuzzy, with the majority of binding events apparently not resulting in the regulation of nearby genes. This raises the question: What distinguishes GR binding events that result in the regulation of gene expression from those that do not? We try to answer this question using a variety of approaches, which advanced our understanding in several ways. First, computational approaches indicate that productive GR binding events are more likely when the three-dimensional organization of the genome brings such regions proximal to the promoter of genes.

Second, using massively parallel reporter assays, we determined the regulatory potential of tens of thousands of GR-bound regions to test if only a subset has an intrinsic regulatory potential. Finally, using genome-editing approaches, we identify single GR-bound recognition sequences that contribute to the GR-dependent regulation of genes. By dissecting an enhancer element with multiple GR recognition sequences and by mutating these, alone or in combination, we uncover that the regulatory potential of the enhancer depends on cooperative interactions between multiple GR binding sites. Furthermore, by building synthetic endogenous*cis*-regulatory elements we investigated the minimal sequence requirements of a functional response element and were able to expand the repertoire of GR target genes. Together, these studies yield several insights into the molecular mechanisms that discriminate â€œproductiveâ€ GR interactions with the genome resulting in gene regulation from â€œnon-productiveâ€ binding events.

Cas9-Based Strategies to Rescue Deep-Intronic Mutations in ABCA4

Conference: International Conference on CRISPR Technologies

Year: 2019

Proceeding: 3rd International Conference on CRISPR Technologies

Group: Poster Session

Session: Registered Posters

Authors:

De Angeli, P., University of Tubingen
Kohl, S., University of Tubingen
Reuter, P., University of Tubingen
Wissinger, B., University of Tubingen

Stargardt disease (STGD) is an autosomal recessive inherited retinal disorder (IRD) caused by biallelic mutations in the ABCA4 gene. IRDs and STGD caused by mutations in ABCA4 are one of the most common forms with an estimated incidence of 1: 10,000. To date, more than 1000 different mutations have been classified as pathogenic. In a considerable number of STGD cases, deep-intronic mutations have been shown to be causative as they result in mRNA mis-splicing. These mRNA-splicing defects lead to either pseudoexon formation or exon skipping which in turn disrupt the normal sequence of codons by frameshift mutations or premature stop codon insertion. As most sequences encompassing deep-intronic mutations have no essential regulatory function, Cas9-based editing exploiting either non-homologous DNA end joining (NHEJ), or Cas9 base editors themselves, offer a potential strategy to rescue deep-intronic mutations. In this project, we apply, assess and compare several Cas9-based approaches to rescue such splicing defects. Additionally, in order to easily assess the different strategies, we also establish fluorescence-reporter vectors which enable to sense the rescue of the deep-intronic mutations back to correct splicing. Preliminary experiments have shown promising results supporting the efficacy and scalability of the tested Cas9-based strategies which can be potentially implemented to target nearly any deep-intronic mutations across the genome.

Co-Expressing the S. Pyogenes Cas9 and Its Tracrrna Induces Potent, Strain-Specific Cytotoxicity in Lactobacilli

CRISPR TECHNOLOGIES CONFERENCE

Conference: International Conference on CRISPR Technologies

Year: 2019

Proceeding: 3rd International Conference on CRISPR Technologies

Group: Poster Session

Session: Registered Posters

Authors:

Vento, J., North Carolina State University
Beisel, C., Helmholtz Institute for RNA-based Infection Research

The Cas9 nuclease from *Streptococcus pyogenes* (SpCas9) has proven to be an invaluable tool for genome editing and gene regulation in bacteria. However, accumulating examples have reported cytotoxicity of expressing Cas9 in some bacteria, limiting its use for genome editing. The underlying mechanism of cytotoxicity and how it can be circumvented remains poorly understood. Here, we report a distinct example in which co-expressing SpCas9 and its trans-activating CRISPR RNA (tracrRNA) is heavily cytotoxic in the industrial bacterium *Lactobacillus paracasei.* We found that transforming a plasmid containing SpCas9 and a tracrRNA resulted in very few colonies in this strain compared to transforming a plasmid containing SpCas9 or the tracrRNA alone. Conversely, this reduction in colonies was not observed when transforming the same plasmid into strains of *Lactobacillus plantarum* and *Lactobacillus rhamnosus*. Interestingly, adding a non-targeting crRNA to the plasmid significantly reduced cytotoxicity in *L. paracasei.* We are currently determining the underlying mechanism by elucidating modes of escape from cytotoxicity, determining the role of DNA nicking and cleavage activity, and identifying putative targets of Cas9. The results of this work are expected to help elucidate the basis of SpCas9 cytotoxicity in bacteria, and in turn help others improve genome editing across bacteria, including industrial relevant Lactobacilli.

CRISPR Array Reconstruction Shields Light on Diversity and Dynamics of Escherichia coli Populations

Conference: International Conference on CRISPR Technologies

Year: 2019

Proceeding: 3rd International Conference on CRISPR Technologies

Group: Poster Session

Session: Registered Posters

Authors:

Savitskaya, E., Skolkovo Institute of Science and Technology
Rykachevckiy, A., Skoltech
Muzyukina, P., Skolkovo Institute of Science and Technology, Skolkovo
Medvedeva, S., Skoltech
Severinov, K., Skolkovo Institute of Science and Technology

Genotyping of *Escherichia coli* isolates is indispensable for clinical diagnostics and epidemiological studies. MLVA (Multiple locus variable number tandem repeat), SNP (single nucleotide polymorphism) and antibiotic resistance spectrum analyzes are not sufficient to meet the needs for fast and discriminative approach. The diversity of CRISPR arrays may constitute a basis for *E. coli* genotyping. Here, we develop an approach based on machine learning. It uses short fragments of CRISPR arrays amplified with primers complementary to CRISPR repeats to reconstruct whole length arrays. The procedure does not need individual clone isolation and is applicable to complex microbiotas. In addition, *E. coli* clonal diversity and dynamics in intestinal samples revealed with the new approach will be discussed.

CRISPR/Cas9 meets recombineering in P. putida KT2440

Conference: International Conference on CRISPR Technologies

Year: 2019

Proceeding: 3rd International Conference on CRISPR Technologies

Group: Poster Session

Session: Registered Posters

Authors:

Aparicio, T., Centro Nacional de Biotecnolog□ (CNB-CSIC)
de Lorenzo, V., Centro Nacional de Biotecnolog□ (CNB-CSIC)
Martanez-Garcaa, E., Centro Nacional de Biotecnolog□ (CNB-CSIC)

The implementation of single-stranded DNA recombineering technology in *Pseudomonas putida* KT2440 has been proven as a rapid and reliable method for genome editing of this platform strain(1). A drawback of this methodology is the low frequency at which the desired changes are introduced in the genome, making it difficult to isolate mutants displaying a non-selectable phenotype. The adoption of CRISPR/Cas9 is an efficient way to counterselect wild-type cells that have not undergone the genetic modifications introduced by ssDNA recombinering, so the combination of both methods dramatically improves the output of this technology, boosting the genome editing capabilities available for *P. putida* KT2440. In this work we report the incorporation of the functional modules for recombineering (ssr recombinase) and CRISPR/Cas9 (Cas9, tracrRNA and the CRISPR array) into pSEVA plasmids to generate a platform that allows a suite of genomic changes ranging form single nucleotide substitutions to large genomic deletions, including the simultaneous deletion of two independent genomic locus(2). An upgrade of the system is also presented, featuring: i) the sevarization of the functional modules ii) a cycling method to sequentially perform deletions without curing the working plasmids. The latest attempts to improve this technology will be also presented, namely the condensation of CRISPR/Cas9/recombineering elements in a single plasmid.

CRISPR/Cas9-Mediated Regulation of GAS5-Encoded Box-C/D Small Nucleolar RNAs

CRISPR TECHNOLOGIES CONFERENCE

Conference: International Conference on CRISPR Technologies

Year: 2019

Proceeding: 3rd International Conference on CRISPR Technologies

Group: Poster Session

Session: Registered Posters

Authors:

Matveeva, A., Institute of Chemical Biology and Fundamental Medicine SB RAS
Julia, F., Institute of Chemical Biology and Fundamental Medicine SB RAS
Evgenii, Z., Institute of Chemical Biology and Fundamental Medicine, SB RAS
Grigory, S., Institute of Chemical Biology and Fundamental Medicine, SB RAS
Evgenia, B., Institute of Chemical Biology and Fundamental Medicine SB RAS
Sergey, M., 3Institute of Fundamental Medicine and Biology, Kazan Federal University
Raihan, S. M., Institute of Fundamental Medicine and Biology, Kazan Federal University
Tatiana, G., Institute of Fundamental Medicine and Biology, Kazan Federal University
Dmitry, S., Institute of Chemical Biology and Fundamental Medicine SB RAS
Valentin, V., ICBFM SB RAS

CRISPR technologies are widely used for a knockout of protein-coding genes or genes encoding long non-coding RNAs (lncRNAs), providing an ability to study and regulate various processes in human cells. However, small non-coding RNAs are a rare target for genome editing. Small nucleolar RNAs (snoRNAs) participate in post-transcriptional modification of ribosomal RNA and guide site-specific 2â€™-O-methylation of target rRNA nucleotides. Their functioning is possible due to the conservative elements in their structure. Objects of the current study are box-C/D-snoRNAs, encoded in introns of the lncRNA GAS5 (Growth Arrest Specific 5) gene. Location of PAM motifs nearby the conserved structural elements allowed obtaining 293FTderived single cell clones with mutations in some SNORDs. Cell lines were characterized for mutations in the selected *GAS5* regions, target RNA level was estimated as well as the 2â€™-O-methylation level of corresponding rRNA nucleotides. We were able to obtain lines with the suppressed activity of individual box-C/D small nucleolar RNAs.

Transcriptome of the modified human cell lines was examined using standard RNA-Seq analysis of polyA fraction as well as small RNAs, including small nuclear RNAs, micro-RNAs and snoRNAs. Our findings confirmed the selectivity of CRISPR/Cas9 editing activity regarding snoRNAs of interest. Analysis of the host gene RNA structure demonstrated partial alterations in mature GAS5 RNA, namely, exon skipping and formation of additional non-canonical exon-exon junctions, only for single obtained cell lines.

Thus, the possibility of obtaining viable cell lines with the suppressed activity of a single snoRNA was demonstrated for studying non-canonical functions of individual snoRNAs, as well as their role in gene expression regulation, post-transcriptional maturation of rRNAs and ribosome assembly.

This work was supported by the RFBR grant â„– 18-29-07073.

CRISPR-Cas9 Mediated Knockout of Host Restriction Factors Allows Increasing Influenza A Virus Replication in Human Cell Line

Conference: International Conference on CRISPR Technologies

Year: 2019

Proceeding: 3rd International Conference on CRISPR Technologies

Group: Poster Session

Session: Registered Posters

Authors:

Grigory, S., Institute of Chemical Biology and Fundamental Medicine, SB RAS
Sergeeva, M., Institute of Chemical Biology and Fundamental Medicine
Vasilieva, A., Institute of Chemical Biology and Fundamental Medicine
Eschenko, N., Institute of Chemical Biology and Fundamental Medicine
Vasiliev, K., Smorodintsev Research Institute of Influenza
Medvedev, S., Federal Research Centre "Institute of Cytology and Genetics
Malakhova, A., Federal Research Centre "Institute of Cytology and Genetics
Evgenii, Z., Institute of Chemical Biology and Fundamental Medicine, SB RAS
Richter, V., Institute of Chemical Biology and Fundamental Medicine
Komissarov, A., Institute of Chemical Biology and Fundamental Medicine

The aim of presented work was exploring the CRISPR-Cas9 mediated knockout of several genes as a way to overcome the host restriction barrier and generate human cell line permissive to influenza infection. We selected some factors including AnxA6 and IRF7 which were earlier described as antivirus system components in human cells. By CRISPR-Cas9 genome editing we obtained several clones of 293FT cells with knockout of these genes. The suppression of target gene expression levels was confirmed with RT-PCR, western-blotting and RNA-Seq analysis.

We compared replication dynamics of several influenza viruses in original and mutated cells in growth curve experiments. Using TCID50 assay, flow cytometry and immunofluorescence staining we showed that accumulation of influenza A virus in the mutant 293FT-ANXA6-/- and 293FT-IRF7-/- cells significantly exceeded the virus titer in the original 293FT cells. Transcriptome analysis was performed to reveal molecular pathways activated in intact and mutant cells during influenza infection. In addition to the upregulation of antiviral response pathways, RNASeq showed the activation of cholesterol metabolism, cell cycle regulation and chromatin remodeling pathways, as well as the downregulation of RNA modification processes. The individual knockout of host cell factors allows describing their role in virus replication control. We propose that multiple gene knockout could provide the desired level of influenza virus replication in adherent 293FT to use this cell line for influenza virus isolation and characterization.

This work was supported by Russian Science Foundation grant 18-75-10069.

Detection of Recbcd-Generated DNA Products in Escherichia coli Cells Undergoing CRISPR Interference

Conference: International Conference on CRISPR Technologies

Year: 2019

Proceeding: 3rd International Conference on CRISPR Technologies

Group: Poster Session

Session: Registered Posters

Authors:

Shiriaeva, A., Rutgers University
Savitskaya, E., Skolkovo Institute of Science and Technology
Kurilovich, E., Skolkovo Institute of Science and Technology
Metlitskaya, A., Institute of Molecular Genetics, Russian Academy of Sciences
Datsenko, K., Rutgers University
Semenova, E., Rutgers University
Severinov, K., Peter the Great St. Petersburg Polytechnic University

CRISPR interference by the *Escherichia coli* subtype I-E CRISPR-Cas system is an efficient process of target degradation by Cas3 helicase/nuclease. In plasmid-based experiments, CRISPR interference leads to full degradation of the target. However, small size of plasmids doesnâ€™t allow to detect the maximal distance to which degradation can occur. We previously developed a self-targeting system where a single chromosomally located protospacer is recognized by the type I-E CRISPR interference machinery. Using this system we show that CRISPR interference proceeds to both directions, upstream and downstream from the targeted protospacer, leading to degradation of up to 300 kb of genomic DNA.

DNA with a double-stranded gap is a substrate for RecBCD-mediated DNA repair in *E. coli.*

Upon binding to a DNA end, RecBCD helicase/nuclease translocates along DNA until it reaches 8-nt sequence termed Chi-site. The recognition of Chi results in nicking of 3â€™-terminated strand and loading of RecA recombinase on a nascent 3â€™end. Despite decades of studying, enzymatic behavior of RecBCD before it pauses at Chi is not understood. *In vitro* RecBCD cleaves both DNA strands producing fragments with a length substantially varying under different reaction conditions. No RecBCD-generated products have been detected *in vivo* yet. We show here that in the self-targeting system degradation of genomic DNA started by CRISPR interference machinery is further continued by RecBCD complex. Using a high-throughput sequencing approach for detecting short DNA fragments, â€œFragSeq,â€ we characterize RecBCD-generated products and establish how cleavage pattern depends on a strand and position with respect to Chi-sites *in vivo*.

Direct Targeting the Expanded CAG Repeat in Huntington's Diseases Using CRISPR/Cas9

Conference: International Conference on CRISPR Technologies

Year: 2019

Proceeding: 3rd International Conference on CRISPR Technologies

Group: Poster Session

Session: Registered Posters

Authors:

Choi, D., Massachusetts General Hospital/ Harvard Medical School
Lee, J. M., Massachusetts General Hospital/ Harvard Medical School

Huntington's diseases (HD) is a dominantly inherited neurodegenerative disease, caused by an expanded CAG trinucleotide repeat in the first exon of huntingtin gene (*HTT*). The length of the repeat length, which is highly polymorphic in the general population, determine one's lifetime risk of developing characteristic HD symptoms such as involuntary movement, progressive decline of cognitive functions, and various psychiatric changes. People with 36-39 *HTT* CAG repeats show reduced penetrance; *HTT* CAG repeats longer than 39 cause fully penetrant movement disorder and neurological changes. In addition, the size of expanded repeat shows a robust inverse correlation with age at onset, indicating the size of the CAG repeat is the most important determinant of the disease. Therefore, inactivation of the mutation may represent the most efficient approach to delay the disease onset and to ameliorate clinical symptoms of this dominant disease.

Aiming at developing novel HD therapeutic strategies, we tested CRISPR/Cas9 genome editing approaches to directly target the *HTT*CAG repeat to inactivate the disease-causing mutation in HD. We screened various *HTT* CAG repeat-targeting CRISPR gRNAs in order to identify CRISPR strategies with therapeutic potential. Specifically, Mi-seq sequencing analysis was performed to determine targeting efficiencies and allele specificities of CRISPR gRNAs in HEK293 cells and induced pluripotent stem cells (iPSC) derived from HD subjects. Our data supported that directly targeting the *HTT* CAG repeat is feasible in HD, informing therapeutic strategies to permanently inactivate the root cause of the disease using CRISPR system to intervene in HD pathogenesis.

1. The *HTT* CAG-Expansion Mutation Determines Age at Death but Not Duration in Huntington Disease. *Am J Hum Genet*. (2016) 98(2):287-298.
2. Novel allele-specific quantification methods reveal no effects of adult onset CAG repeats on *HTT* mRNA and protein levels. *Hum Mol Genet*. (2017) 26(7):1258-1267.
3. Huntington's disease: CAG genetics expands neurobiology. *Curr Opin Neurobiol*. (1995) 5(5):656-62.

DNA Mimic Protein Ocr from Phage T7 Is a Potent Inhibitor of Brex Defence System

Conference: International Conference on CRISPR Technologies

Year: 2019

Proceeding: 3rd International Conference on CRISPR Technologies

Group: Poster Session

Session: Registered Posters

Authors:

Isaev, A., Skokovo Institute of Science and Technology
Drobiazko, A., Skokovo Institute of Science and Technology
Gordeeva, J., Skokovo Institute of Science and Technology
Severinov, K., Peter the Great St. Petersburg Polytechnic University

Bacteriophage Exclusion (BREX) is a phage defense system found in more than 10% of microbial genomes. In recent work we have shown that BREX from *Escherichia coli* HS provides protection against several unrelated DNA phages and discriminates between self and non-self by methylation of a specific DNA site. The mechanism of defensive action of BREX remains unclear, however. While T7 phage is able to productively infect BREX cells, we observed that a mutant phage without a functional*o.3* gene is subject to BREX exclusion. T7 gene *o.3* codes for the DNA mimic protein Ocr, a well-known inhibitor of type I restriction-modification systems. Expression of the *o.3* gene from a plasmid abolished BREX defense against T7 and other unrelated phages. Pull-down experiments demonstrated direct interaction of Ocr with BREX system methyltransferase BrxX. Yet, the DNA methylation level of cells harboring BREX system was not diminished in the presence of Ocr, suggesting that the inhibition of BREX defence occurs at a stage other than DNA methylation.

The study was supported by the RFBR grants â„–18-34-00845 mol_a and â„–19-34-90160.

Endogenous Tagging Strategies for Accelerated Target Validation and Detection

Conference: International Conference on CRISPR Technologies

Year: 2019

Proceeding: 3rd International Conference on CRISPR Technologies

Group: Poster Session

Session: Registered Posters

Authors:

Ciotta, G., AstraZeneca
Hock, A., AstraZeneca
Carter, N., AstraZeneca
Cuomo, E., AstraZeneca

CRISPR/Cas9 system has facilitated precise insertions in the genome of a cell to monitor endogenous activities of genes. This method has advantages compared to target overexpression under an exogenous promoter which may cause numerous artifacts including ectopic sub-cellular localizations and erroneous formation of protein complexes. As opposite to an irreversible gene knockout, precise temporal and spatial control of protein expression enables a detailed analysis of the cellular phenotype or the investigation of essential genes. PROTACs (Proteolysis-Targeting Chimera) may be combined with CRISPR to establish a universal approach to generate cell models for Target Validation (TV) where the target is fused to both a tag sensitive to PROTAC degradation (such as the HaloTag), and a split nanoluciferase.

We have developed a rapid workflow for cell line engineering based on an endogenous tagging strategy utilizing long single-stranded DNA (ssDNA; 2000-4000 nt in length) HDR donors. Knockin pools or clones can then be used in protein degradation assays via PROTAC to interrogate the protein function. The decrease of the target protein and the speed of the protein degradation can be readily monitored by luciferase activity.

Establishing a CRISPR/Cas9 Based System Designed to Explore the Similarities and Differences between Lysyl Oxidases

CRISPR TECHNOLOGIES CONFERENCE

Conference: International Conference on CRISPR Technologies

Year: 2019

Proceeding: 3rd International Conference on CRISPR Technologies

Group: Poster Session

Session: Registered Posters

Authors:

Nir Zvi, I., Technion
Liburkin Dan, T., Technion
Kessler, O., Technion
Neufeld, G., Technion

Lysyl-oxidases catalyze the covalent cross-linking of collagens and contribute to the stabilization of extracellular matrixes. The LOX family contains five genes (LOX, LOXL1-4). They are characterized by a highly conserved C-terminal catalytic domain that includes a copper binding domain. Several lysyl-oxidase family members such as LOX and LOXL2 were found to promote tumor metastasis, and tumor angiogenesis. They were found to promote tumor progression by diverse mechanisms that include induction of epithelial to mesenchymal transition (EMT) and extracellular matrix remodeling that results in enhanced extracellular matrix stiffness resulting in enhanced tumor cell invasiveness and tumor metastasis. However, it is unclear if different lysyl-oxidases all promote tumor progression or if differences between individual lysyl-oxidases result in different effects of individual lysyl-oxidases on tumor progression.

We created novel MDA-MB-231 and LM2-4 cell lines in which we have knocked-out the five LOX family genes, using the CRISPR/Cas9 method. This resulted in the generation of viable

LM2-4 and MDA-MB-231 cell lines that do not express any lysyl-oxidases. These cells displayed significant inhibition of invasive ability and were unable to form colonies in soft agar. To identify similarities and differences between the different lysyl-oxidases we re-expressed in these cells cDNAs encoding individual lysyl-oxidases and we then examined the effects of the re-expression on gene expression profiles using deep sequencing and mass spectrometry. This resulted in the identification of several genes whose expression was regulated by individual lysyl-oxidases.

Evaluation of CRISPR Spacers Diversity in Clostridium Difficile Clinical Isolates

Conference: International Conference on CRISPR Technologies

Year: 2019

Proceeding: 3rd International Conference on CRISPR Technologies

Group: Poster Session

Session: Registered Posters

Authors:

Muzyukina, P., Skolkovo Institute of Science and Technology, Skolkovo
Severinov, K., Skolkovo Institute of Science and Technology
Savitskaya, E., Skolkovo Institute of Science and Technology

Antibiotic-associated colitis caused by enteropathogen *Clostridium difficile* represents a great challenge through multiple antibiotic resistance these bacteria possess[1] . Recently, bacteriophages have been reported as a promising alternative to antimicrobial agents[2]· Regarding*C. difficile* infection, the phage treatment has the potential to become a biologically solid and highly efficient approach[3] , though its usage might be limited by active CRISPR-Cas defense systems of this pathogen[4] . Along with *cas* genes, CRISPR-Cas system of *C. difficile* contains up to ten CRISPR arrays. By targeting complementary sequences in mobile genomic elements, such as phages and plasmids, spacers provide immunity to host cells and their analysis can reveal a history of host-phage interactions and inform on the choice of therapeutic phages. Our study aims to develop a straightforward analysis of CRISPR spacers of *C. difficile* in clinical isolates and whole communities. The approach relies on PCR with oligonucleotides complementary to *C. difficile* CRISPR repeats, followed by high throughput sequencing and bioinformatics reconstruction of whole arrays. This procedure should provide information necessary for fast assessment of a phage-sensitivity of *C. difficile* strains in patientsâ€™ samples. Based on this information individual bacteriophage cocktail treatment can be designed.

1. Peng Z, Jin D, Kim HB, Stratton CW, Wu B, Tang YW, Sun X. Update on Antimicrobial Resistance in Clostridium difficile: Resistance Mechanisms and Antimicrobial Susceptibility Testing. Clin. Microbiol 2017; 55.

2. Abedon ST, GarcÃa P, Mullany P, Aminov R. Phage Therapy: Past, Present and Future. Frontiers in Microbiology 2017; 8.

3. Meader E, Mayer M, Gasson M, Steverding D, Carding S, Narbad A.Bacteriophage treatment significantly reduces viable Clostridium difficile and prevents toxin production in an in vitro model system. Anaerobe 2010; 16: 549â€“554

4. Soutourina OA, Monot M, Boudry P, Saujet L et al. Genome-wide identification of regulatory RNAs in the human pathogen Clostridium difficile. PLoS Genet 2013; 9(5):e1003493.

Extensive Self-Targeting By CRISPR-Cas Systems in the Plant Pathogen Xanthomonas Albilineans

Conference: International Conference on CRISPR Technologies

Year: 2019

Proceeding: 3rd International Conference on CRISPR Technologies

Group: Poster Session

Session: Registered Posters

Authors:

Wimmer, F., Helmholtz Institute for RNA-based Infection Research (HIRI), Helmholtz-Centre for Infection Research (HZI), 97080 Wurzburg, Germany
Beisel, C., Helmholtz Institute for RNA-based Infection Research

CRISPR-Cas systems are RNA-directed adaptive immune systems in bacteria and archaea. Genetic information of invaders is collected and incorporated as spacers in CRISPR arrays. As targeting relies on spacers, self-targeting spacers are expected to lead to autoimmunity and cell death. Nevertheless, there exist examples of strains with genome-targeting spacers. Xanthomonas albilineans is a unique example with over 20 self-targeting spacers, the largest content reported to date. The bacterium also harbors two CRISPR-Cas systems, a type I-C and a type I-F system and all cas genes required for these systems are encoded within X. albilineans genome. Type I systems encode a multiâ€‘protein effector complex called Cascade that is sufficient for target binding and, after recruiting Cas3, degrades its target. We used a cell-free transcription-translation system (TXTL) to characterize X. albilineansCRISPR-Cas systems and to address the function of these genome-targeting spacers.

We found that both systems could yield target DNA binding as well as cleavage activity. Furthermore, 5â€™CC and 5â€™TTC for I-F and I-C system, respectively, are functional

PAMs. These PAMs are also associated with genome-targeting spacers, suggesting extensive self-targeting. Anti-CRISPR proteins encoded in X. albilineans genome could prevent the bacterium from lethal self-targeting. Supporting this hypothesis, we could identify ~15 putative anti-CRISPR proteins encoded within X. albilineans genome or plasmids. These proteins could serve as regulators and prevent cell-death. With this, a new alternative function for CRISPR-Cas systems as gene-regulating system could be discovered and new information about the evolution of CRISPR-Cas systems can be gathered.

Generation and Validation of Human Pluripotent Stem Cell Disease Models Using CRISPR/Cas9 Genome Engineering

Conference: International Conference on CRISPR Technologies

Year: 2019

Proceeding: 3rd International Conference on CRISPR Technologies

Group: Poster Session

Session: Registered Posters

Authors:

Turan, S., Department of Stem Cell Biology, Friedrich-Alexander-Universitat Erlangen-Narnberg (FAU), 91054 Erlangen, Germany
Kavyanifar, A., Institute of Biochemistry, Friedrich-Alexander-Universitiut Erlangen-Nurnberg, 91054 Erlangen, Germany (FAU)
Wend, H., Department of Stem Cell Biology, Friedrich-Alexander-Universit□ Erlangen-Nurnberg (FAU), 91054 Erlangen, Germany
Pozner, T., Department of Stem Cell Biology, Friedrich-Alexander-Universit□ Erlangen-Nurnberg (FAU), 91054 Erlangen, Germany
Loskarn, S., Department of Stem Cell Biology, Friedrich-Alexander-Universit□ Erlangen-Nurnberg (FAU), 91054 Erlangen, Germany
Kaindl, J., Department of Stem Cell Biology, Friedrich-Alexander-Universit□ Erlangen-Nurnberg (FAU), 91054 Erlangen, Germany
Lanfer, J., Department of Stem Cell Biology, Friedrich-Alexander-Universit□ Erlangen-Nurnberg (FAU), 91054 Erlangen, Germany
Rizo, T., Department of Stem Cell Biology, Friedrich-Alexander-Universit□ Erlangen-Nurnberg (FAU), 91054 Erlangen, Germany
Riedlberger, J., Department of Stem Cell Biology, Friedrich-Alexander-Universit□ Erlangen-Nurnberg (FAU), 91054 Erlangen, Germany

Winner, B., Department of Stem Cell Biology, Friedrich-Alexander-Universit□ Erlangen-Nurnberg (FAU), 91054 Erlangen, Germany
Lie, D. C., Institute of Biochemistry, Friedrich-Alexander-Universitiut Erlangen-Nurnberg, 91054 Erlangen, Germany (FAU)

CRISPR/Cas9 mediated genome engineering of human pluripotent stem cells enables recapitulation or correction of patient-specific disease mutations allowing modelling of genetic disorders, when patient material is scarce. Disease models serve as a platform to allow detection of human specific disease mechanisms and to elucidate dysfunctional pathways. We developed a CRISPR/Cas9 genome editing protocol for hESCs and hiPSCs to efficiently generate indel based heterozygous or homozygous knockout models. This protocol includes usage of an eGFP tagged eCas9 for FACS-based single cell sorting of nucleofected stem cells to generate pure clonal lines. CRISPR mediated indels are detected by Sanger sequencing and NGS amplicon sequencing. Further genetic quality control guidelines include PCR based off-target analysis and CNV analysis. We were able to efficiently generate functional knockout lines for several genes, which play a central role for neurodevelopmental disorders such as intellectually disability-associated Coffin-Siris syndrome (SOX11), Pitt Hopkins syndrome (TCF4), and genes that were found critical in the pathophysiology of neurodegenerative disorders such as Parkinsonâ€™s disease (SNCA) and hereditary spastic paraplegia (SPG4 or SPG11). Interestingly, lines derived after treatment of some sgRNAs showed higher incident of *de novo* CNVs, despite no signs of off-target mutagenesis.

Currently, we are focusing on steps to optimize the efficiency for generation of corrected patient hiPSC lines by employing homology directed repair (HDR)-based mechanisms for single nucleotide substitutions. This includes testing of small molecule HDR enhancers, different sorting and analyzing strategies. Finally, we are testing novel PCR or donor based methods to generate FLAG or fluorescence marker tagged HEK293T and hPSC reporter lines.

Genome Editing for FVIII and AAT Gene Therapy Cures

Conference: International Conference on CRISPR Technologies

Year: 2019

Proceeding: 3rd International Conference on CRISPR Technologies

Group: Poster Session

Session: Registered Posters

Authors:

Lorincz, R., Washington University in St Louis
Curiel, D., Washington University in St Louis
Dmitriev, I., Washington University in St Louis
Lu, Z. H., Washington University in St Louis

The CRISPR/Cas9 nuclease system is a powerful tool for manipulating genomes by DNA editing. This allows for functional knock-in for gene therapy correction. However, current limits of *in vivo* gene delivery methods have practically restricted clinical application. In this regard, the recombinant adeno-associated viral (AAV) vectors have been mainly applied for *in vivo* CRISPR/Cas9-mediated gene therapy studies. However, native AAV tropism limits the possibility of organ selective gene editing. In addition, AAVâ€™s innate liver tropism is associated with hepatotoxicity. As an anodyne, we explored adenoviral (Ad) vector to achieve effective CRISPR/Cas9 based knock-in. The unique capacity to alter Ad tropism can address both of these limits. For alpha1-antitrypsin (AAT) deficiency lung disease, the augmentation of AAT within the lower respiratory tract is the key biologic mandate, thus we targeted Ad vector to the pulmonary endothelium. This pulmonary targeted Ad approach has been designed to accomplish gene therapy for AAT lung disease in a murine model of this disorder (SERPINA1a-e KO mice). For hemophilia A gene therapy, we have pursued an approach to limit potential hepatotoxicity. Specifically, we have sought a non-hepatic source of the corrective factor VIII

(FVIII). Thus we explored a pan-endothelium delivery so that vascular endothelium can serve this goal. We designed an optimal co-delivery of CRISPR/Cas9 and FVIII therapeutic gene by a myeloid cell-binding Ad that allows effective and selective *in vivo* editing for endothelial target cells to achieve phenotypic correction in hemophilia A murine model. Here, we propose a highly original vectorology to traverse key barriers for efficient *in vivo* genome editing-based gene therapy in order to achieve long term genetic correction of hemophilia A and AAT deficiency.

Harnessing Endogenous Repair Mechanisms for Precise Modification of in Vitrofertilized Bovine Embryos

Conference: International Conference on CRISPR Technologies

Year: 2019

Proceeding: 3rd International Conference on CRISPR Technologies

Group: Poster Session

Session: Registered Posters

Authors:

Owen, J., University of California, Davis
Hennig, S., University of California, Davis
Paulson, E., University of California, Davis
Lin, J., University of California, Davis
Ross, P., University of California, Davis
Van Eenennaam, A., University of California, Davis

The use of genetic editing tools, such as the CRISPR/Cas9 system, could be used to introduce useful health and welfare characteristics such as disease resistance into cattle breeding programs. However, current methods for harnessing these tools for knock-ins or whole gene insertions have been limited to alterations in cell culture followed by somatic cell nuclear transfer (SCNT) cloning. This approach is problematic as SCNT cloning is inefficient, and edits are made in existing germplasm, rather than the next generation. One alternative is CRISPR-mediated modification of in vitrofertilized embryos by direct cytoplasmic injection. Overall, this method has been unsuccessful at achieving knock-ins in cattle embryos due to the low activity of the homologous recombination (HR) pathway in gametes and one-cell zygotes. Previous approaches for attempting gene knock-ins using homology directed repair (HDR) have primarily focused on improving HR function by knocking down the non-homologous end-joining (NHEJ) pathway, or attempting to enhance HR activity. These methods have been largely unsuccessful in embryos as double strand breaks are primarily resolved using the NHEJ pathway. We have developed an approach which utilizes the homology mediated end-joining (MHEJ) pathway to facilitate HDR via direct cytoplasmic injection of in vitro matured bovine oocytes prior to fertilization. Using this method we showed a significantly higher rate of gene knock-in as compared to approaches which utilize the HR pathway (34.0% vs. 0%; $p < 0.0001$). Along with the increased rate of gene integration at the target site, we observed a high level of non-mosaic embryos when using this approach (24.6%). As the end-joining pathway is the primary repair mechanism in gametes, this method harnesses the ability for HDR by direct injection of oocytes, utilizing the MHEJ pathway for targeted gene knock-in of embryos. This approach could enable precise germline editing of the next generation of elite bovine embryos.

In Vivo analysis of CAS1/2-CAS3 Proteins Interaction during Primed Adaptation

Conference: International Conference on CRISPR Technologies

Year: 2019

Proceeding: 3rd International Conference on CRISPR Technologies

Group: Poster Session

Session: Registered Posters

Authors:

Musharova, O., Institute of Molecular Genetics, Russian Academy of Sciences
Klimuk, E., Institute of Molecular Genetics
Savitskaya, E., Skolkovo Institute of Science and Technology
Severinov, K., Skolkovo Institute of Science and Technology

CRISPR-Cas systems are responsible for two different cellular phenomena: CRISPR interference and CRISPR adaptation. CRISPR interference occurs when a protospacer recognized by the spacer segment of CRISPR RNA is destroyed by Cas effectors. CRISPR adaptation is an integration of new spacers into an array, that relays on conserved Cas1-Cas2 protein complex. For Type I CRISPR-Cas system a unique mode of CRISPR adaptation, named primed adaptation, has been described [Datsenko et al., 2012]. In this course the efficiency of new spacer acquisition is stimulated by recognition of target DNA based on the presence of pre-existed spacer. In *Esherichia coli* type I-E system the ribonucleoprotein surveillance Cascade complex containing crRNA, which is the product of CRISPR array transcription and processing, recognizes the foreign DNA [Jore et al., 2011]. This recognition serves as a signal for engaging of Cas3 protein which unwinds and degrades the target DNA [Westra et al., 2012]. The Cas1-Cas3 complex was shown at some extent facilitates primed adaptation [Dillard et al., 2018]. We for the first time performed chromatin immunoprecipitation procedure with Cas3-specificic or Cas1-specific antibodies to characterize DNA fragments associated with those proteins *in vivo*.

Close interplay between CRISPR-interference and spacer integration during primed adaptation was revealed based on shared patterns on DNA fragments obtained with anti-Cas1 and anti-Cas3 antibodies. We also analyzed direct interaction between Cas1 and Cas3 protein *in vivo*. For that propose we obtained*E.coli* strain containing His-tagged Cas1 protein and *cas3* gene deletion. Those cells were transformed with plasmid expressing Cas3 Strep tag and subjected to two-stages chromatography. This study was supported by the Russian Foundation for Basic Research grant 18-34-00048 O.M.

Modular, One-Pot Assembly of CRISPR Arrays Enables Library Generation, Multi-Nuclease Targeting and Reveals Factors Influencing CRISPR RNA Biogenesis

Conference: International Conference on CRISPR Technologies

Year: 2019

Proceeding: 3rd International Conference on CRISPR Technologies

Group: Poster Session

Session: Registered Posters

Authors:

Liao, C., Helmholtz Institute for RNA-based Infection Research
Ttofali, F., North Carolina State University
Slotkowski, R. A., North Carolina State University
Denny, S. R., Columbia University
Leenay, R. T., North Carolina State University
Keung, A., North Carolina State University
Beisel, C., Helmholtz Institute for RNA-based Infection Research

CRISPR-Cas systems inherently multiplex through CRISPR arrays--whether to defend against different invaders or mediate multi-target editing, regulation, imaging, or sensing. However, arrays remain difficult to generate due to their reoccurring repeat sequences. Here, we report a modular, one-pot scheme called CRATES to construct CRISPR arrays and array libraries. CRATES allows assembly of repeat-spacer subunits using defined assembly junctions within the trimmed portion of spacers. Using CRATES, we constructed arrays for the single-effector nucleases Cas9, Cas12a, and Cas13a that mediated multiplexed DNA/RNA cleavage and gene regulation in cell-free systems, bacteria, and yeast. CRATES further allowed the one-pot construction of array libraries and â€œcompositeâ€ arrays utilized by multiple Cas nucleases. Finally, array characterization revealed processing of extraneous CRISPR RNAs from Cas12a terminal repeats and sequence- and context-dependent loss of crRNA-directed nuclease activity via global RNA structure formation. CRATES thus can facilitate diverse multiplexing applications and help identify factors impacting crRNA biogenesis.

Natural Diversity of CRISPR Spacers

Conference: International Conference on CRISPR Technologies

Year: 2019

Proceeding: 3rd International Conference on CRISPR Technologies

Group: Poster Session

Session: Registered Posters

Authors:

Medvedeva, S., Skoltech
Severinov, K., Peter the Great St. Petersburg Polytechnic University
Liu, Y., Institute Pasteur
Krupovic, M., Institute Pasteur

Hyperthermophilic archaea of the order Sulfolobales harbor some of the most complex among the studied CRISPR-Cas systems: most of the genomes contain several CRISPR arrays with different CRISPR repeats, several adaptation modules for acquisition of new spacers into CRISPR arrays and several type I and type III interference modules that degrade the DNA and/or RNA molecules of encountered mobile genetic elements (MGEs). The sequence of each CRISPR spacer and its position in the array, respectively, provide information about the encountered MGEs and the order of their interaction with the host.

Optimization and Modeling of Crrna Design for Pathogen Detection By Cas13 Enzymes

Conference: International Conference on CRISPR Technologies

Year: 2019

Proceeding: 3rd International Conference on CRISPR Technologies

Group: Poster Session

Session: Registered Posters

Authors:

Schultzhaus, Z., National Research Council
Wang, Z., Naval Research Laboratory
Stenger, D., Naval Research Laboratory

Type VI Cas proteins possess non-discriminatory endonuclease activity that is triggered by the binding of the CRISPR ribonucleoprotein complex to target sequences. This feature has been coopted to enable customizable detection of nucleic acids through coupling nuclease activity to molecular beacons, providing the potential to create made-to-order, point-of-care diagnostics with strain discrimination capabilities. However, the speed of development, the sensitivity, and the specificity of this assay is dependent upon, in part, features of the crRNA, as well as the Cas protein used. We are interested in improving protocols for crRNA development that minimizes cost and development time. Toward this goal we have synthesized and tested several crRNAs tiled against loci in*Yersinia* species, for use with Cas13a proteins from *Leptotrichia buccalis*and *Leptotrichia wadei*, and determined features (including GC content and predicted, ancillary secondary structure) of the crRNAs that provided the most robust, specific readout using fluorescent small RNA reporters. Our results indicate that false positives are a concern with Cas13a orthologs from certain species, and that signal intensity varies significantly between different but closely related crRNAs. However, certain conditions and crRNAs can provide discriminatory capabilities down to the level of 1 and 2 nucleotides, which is sufficient to distinguish many closely related bacterial strains.

Outcomes of Genome Targeting in Bacteria Vary between CRISPR Nucleases

Conference: International Conference on CRISPR Technologies

Year: 2019

Proceeding: 3rd International Conference on CRISPR Technologies

Group: Poster Session

Session: Registered Posters

Authors:

Vialetto, E., Helmholtz Institute for RNA-based Infection Research
Beisel, C., Helmholtz Institute for RNA-based Infection Research
Yu, Y., Helmholtz Institute for RNA-based Infection Research
Collias, D., North Carolina State University
Barquist, L., University of Wurzburg

The discovery of CRISPR-Cas immune systems in bacteria and archaea has led to many applications based on programmable DNA and RNA targeting with each systemâ€™s RNA-guided nucleases. Based on the observation that genome targeting in bacteria leads to cell death, these nucleases have been used to eradicate cells that failed to undergo recombineering as part of gene editing or exhibit virulence or pathogenicity as part of programmable antimicrobials. Previous studies have used individual single-effector nucleases in single strains for these purposes. However, how these nucleases compare across strains remains to be explored. In this study, we compared three single-effector nucleases with different targeting mechanisms, type II-A Cas9, type V-A Cas12a and type VI-A Cas13a, in different bacteria. We initially tested the genome-targeting activity of these nucleases in Escherichia coliMG1655. We found that each nuclease yielded widely varying reductions in the transformation efficiency of a targeting versus non-targeting guide RNA plasmid, with Cas12a yielding the largest reduction. We also identified different mechanisms of escape from targeting, including repair of cleaved DNA by RecA for

Cas9 and delayed cell death or recombination within the guide RNA for Cas12a. Finally, we identified distinct means to counter the survival mechanisms, thereby enhancing cell death and reducing escape. These results show that the outcomes of genome targeting with Cas single-effector nucleases in bacteria can vary between nucleases, with implications for the use of CRISPR for genome editing and programmable antimicrobials in bacteria.

Possible Targets for Heterochromatin to Euchromatin Transition of Genes Using CRISPR/dCas9-Mediated Epigenome Editing

Conference: International Conference on CRISPR Technologies

Year: 2019

Proceeding: 3rd International Conference on CRISPR Technologies

Group: Poster Session

Session: Registered Posters

Authors:

Keshavarz Farajkhah, K., Shahid Beheshti University of Medical Sciences, School of Pharmacy, Tehran, Iran

Among different epigenetic mechanisms affecting transcription, lots of efforts have been made to alter the state of DNA methylation and also activation/repression of gene expression through targeting cis-regulatory elements using nuclease deficient Cas9 (dCas9). Although chromatin modification by CRISPR/Cas9 technology allocates a significant part of researches in the field of epigenome editing, researchers have focused only on deactivation of genes by changing the locus state of genes from euchromatin to heterochromatin (1). Heterochromatin/euchromatin status of chromosome is dynamic and Environmental factors may lead to transition of loci state from euchromatin to heterochromatin and somehow this change may remain through the rest of cell life. Heterochromatin protein HP1a and histone H3K9 methyltransferase are the key elements in position-effect variegation (PEV) and heterochromatin formation and as a result, juxtaposed genes of euchromatin with heterochromatin (2). Histone methylation is not irreversible and several histone demethylases are expressed in cells. HP1a contributes not only in gene silencing but also in activation of genes. This protein can bind to different proteins such as histone methyl-transferases and demethylases and since the effect of histone methylation on gene expression depends on the position of methylated lysine(3), engineering the methylation status of adjacent histones to a specific gene by CRISPR/dCas9 would be a promising idea. In this review, I want to study the feasibility of using HP1 and various histone methyl-transferases and demethylases for CRISPR/dCas9-mediated epigenome editing and discuss the advantages and possible obstacles.

References:

- Pulecio, J, et al., 2017. CRISPR/Cas9-based engineering of the epigenome. *Cell Stem Cell*, *21*(4), pp.431-447.
- Elgin, S.C. and Reuter, G., 2013. Position-effect variegation, heterochromatin formation, and gene silencing in Drosophila. *Cold Spring Harbor perspectives in biology*, *5*(8), p.a017780.
- Kwon, S.H. and Workman, J.L., 2011. The changing faces of HP1: from heterochromatin formation and gene silencing to euchromatic gene expression. *Bioessays*, *33*(4), pp.280-289.

Precision gene-engineering meets cancer immunotherapy: Targeted insertion of chimeric antigen receptors into the T-cell genome using a novel, virus-free CRISPR/Cas9 knock-in strategy

Conference: International Conference on CRISPR Technologies

Year: 2019

Proceeding: 3rd International Conference on CRISPR Technologies

Group: Poster Session

Session: Registered Posters

Authors:

Weber, J., Universitutsklinikum Wurzburg, Medizinische Klinik und Poliklinik II

Introduction Immunotherapy with gene-engineered CAR-T-cells is a transformative cancer treatment. In conventional CAR-T-cell products, gene-transfer is accomplished with viral vectors that are associated with genotoxicity, a risk for insertional mutagenesis, and substantial effort and expense of vector production and handling. To overcome these limitations, we established virus-free targeted insertion to direct the CAR-transgene into defined genomic loci in human T-cells.

Results We validated CRISPR gRNAs targeting the endogenous T-cell-receptor (TCR), and generated CAR knock-in T-cells. We confirmed TCR knock-down and CAR knock-in by flow cytometry and on the genomic level. Notably, CAR knock-in T-cells displayed i. lower and more homogeneous CAR surface expression, ii. activation-dependent CAR regulation, and iii. reduced levels of baseline activation and exhaustion marker expression compared to T-cells with random genomic insertion.

CAR knock-in T-cells conferred specific cytotoxicity, cytokine secretion, and outperformed conventional (random insertion) CAR-T-cells in antigen-dependent proliferation. These data were confirmed with several CAR-constructs of relevance in hematologic (CD19) and solid tumor indications (ROR1/2). Preliminary data also suggest that targeted CAR insertion into the TCR locus is advantageous to the PD1 locus, likely because it enables T-cells to regulate CAR expression physiologically, in sync with their activation status. Accordingly, we consistently found improved viability and lower rates of AICD in knock-in CAR-T-cells after a tumor challenge.

Outlook We are presenting virus-free targeted CAR insertion into a genomic safe harbor as a strategy to augment safety and performance of CAR-T-cell therapy; and consider precision genomic engineering as a key enabling technology in the field of ATMPs.

Specificity of Prespacer Choice during CRISPR Adaptation

Conference: International Conference on CRISPR Technologies

Year: 2019

Proceeding: 3rd International Conference on CRISPR Technologies

Group: Poster Session

Session: Registered Posters

Authors:

Vyhovskyi, D., Skoltech
Savitskaya, E., Skolkovo Institute of Science and Technology
Severinov, K., Skolkovo Institute of Science and Technology

CRISPR arrays comprising unique spacers separated by identical repeats ensure prokaryotic immunity through specific targeting of foreign nucleic acids complementary to spacers. Spacers are acquired into arrays as fragments of foreign DNA (prespacers) [1]. Different fragments are acquired into CRISPR arrays with widely different efficiencies but factors responsible for acquisition efficiency are not fully known.

In this work, we studied the effects of prespacer sequence alterations on the acquisition efficiency by the type I-E CRISPR-Cas system using library-based approaches as well as by introducing specific mutations in prespacer sequences. Both naïve and primed adaptation was investigated.

Our work revealed structures and motives that were depleted from spacers acquired with a high frequency. Most prominently, the presence of internal consensus PAM trinucleotides in prespacers had a strong negative effect on spacer acquisition efficiency during primed but not naïve adaptation [2]. Similar trends of avoidance of cognate PAMs were observed when spacers acquired by different CRISPR-Cas systems (type I-B, III-B, I-F, II-A) were analyzed.

[1] Savitskaya E, Semenova E, Dedkov V, Metlitskaya A, Severinov K. High-throughput analysis of type IE CRISPR-Cas spacer acquisition in E. coli. RNA biology. 2013; 10(5): 716-725.

[2] Musharova O, Vyhovskyi D, Medvedeva S, Guzina J, Zhitnyuk Y, Djordjevic M, Savitskaya E. Avoidance of Trinucleotide Corresponding to Consensus Protospacer Adjacent Motif Controls the Efficiency of Prespacer Selection during Primed Adaptation. mBio. 2018; 9(6), e02169-18.

SuCms1 is a type V CRISPR nuclease variant with a broad PAM preference and tolerance to anti-CRISPR proteins

Conference: International Conference on CRISPR Technologies

Year: 2019

Proceeding: 3rd International Conference on CRISPR Technologies

Group: Poster Session

Session: Registered Posters

Authors:

Dmytrenko, O., Helmholtz Institute

Class 2 type V CRISPR-Cas systems are characterized by a single effector nuclease which requires guide RNA for activity, making them attractive genome-editing tools. A distinct group of type V effector variants, termed Cms1, has recently been identified in metagenomic datasets. Their size and domain architecture set them apart from other known type V nucleases, suggesting likely functional differences. Many of the more than 40 predicted putative Cms1 nucleases exhibit editing activity in plants. However, catalytic properties and biotechnological potential of these effectors remain largely unexplored. In this study we characterized the SuCms1 nuclease identified in a metagenomic sequence of an uncultured chemolithoautotrophic epsilonproteobacterium, Sulfuricurvum spp., from the Frasassi Cave, Italy. We carried out our assessment of the fundamental properties of SuCms1 in Escherichia coli BL21-AI heterologous host. SuCms1 exhibited two nuclease activities, mediating CRISPR-RNA biogenesis and catalyzing RNA-dependent interference by cleaving dsDNA targets. This cleavage was dependent on the three RuvC domains, similar to Cas12a. However, unlike any know type V nuclease, SuCms1 exhibited a broad PAM preference, recognizing approximately half of the nucleotide combinations in a 5N library. Furthermore, the role of the -5 position, which is not part of the Cas12a consensus PAM, was context dependent in the case of SuCms1. Finally, we showed that known type V anti-CRISPR (AcrVA) proteins did not inhibit SuCms1. Particularly notable is the lack of inhibition by AcrVA5 which inactivates Cas12a by acetylating a conserved PAM recognition lysine residue also found in SuCms1. These data suggest that the PAM recognition mechanism of SuCms1 is entirely different from that of Cas12a, potentially relying on a combination of sequence identities and associated structural DNA features. This work helps establish SuCms1 as a unique variant among type V CRISPR-nucleases and informs efforts of adopting Cms1 effectors for biotechnological applications.

The Impact of Cellular Genome Maintenance Proteins on Primed CRISPR Adaptation By the Escherichia coli Type I-E CRISPR-Cas System

Conference: International Conference on CRISPR Technologies

Year: 2019

Proceeding: 3rd International Conference on CRISPR Technologies

Group: Poster Session

Session: Registered Posters

Authors:

Kurilovich, E., Skolkovo Institute of Science and Technology
Shiriaeva, A., Rutgers University
Metlitskaya, A., Institute of Molecular Genetics, Russian Academy of Sciences
Ivan?i?-Ba?e, I., Faculty of Science, University of Zagreb, Croatia
Savitskaya, E., Skolkovo Institute of Science and Technology
Severinov, K., Peter the Great St. Petersburg Polytechnic University

Primed CRISPR adaptation by the type I-E CRISPR-Cas system of*Escherichia coli* is coupled with CRISPR interference and directed by the protospacer targeting. During this process, short DNA fragments from the priming area are incorporated into the CRISPR array to become new spacers, with efficiency of selection gradually falling with a growing distance from the protospacer. However, the exact mechanism of spacer choice and insertion and the participants of these reactions *in vivo* are not yet fully determined. In this study, we use previously described self-targeting model to establish the impact of several proteins providing bacterial genome stability in primed CRISPR adaptation. By introducing single or double knockout mutations of the corresponding genes and performing high-throughput analysis of acquired spacers, we compare the efficiency and patterns of CRISPR adaptation. Based on the results, the distinct roles of RecBCD, SbcCD, SbcB and RecJ proteins in different steps of primed CRISPR adaptation are proposed, suggesting these proteins to alter the processivity of priming complex or pre-spacer maturation and insertion.

Utilizing CRISPR/Cas Technology for Point-of-Care Diagnosis

Conference: International Conference on CRISPR Technologies

Year: 2019

Proceeding: 3rd International Conference on CRISPR Technologies

Group: Poster Session

Session: Registered Posters

Authors:

Franch, O., Delft University of Technology
Bengtson, M., Technical University Delft
Bharadwaj, M., Delft University of Technology
Torre, J. V. D., Delft University of Technology
Dekker, C., Kavli Institute of Nanoscience, Delft. Delft University of Technology

CRIPSR/Cas complexes possess an unparalleled ability to find specific nucleotide sequences in vast amounts of genomic material. We utilize this capability to develop a diagnostic test that probes for the presence of a pathogenâ€™s DNA in body fluids.

We are developing an isothermal DNA-detection point-of-care diagnostic test that is equipment free and has a robust temperature range of operation. The DNA of a pathogen is detected with a CRISPR/Cas9 system using a guide RNA that is selected for repetitive sequences in the pathogenâ€™s genome. The CRISPR/Cas9 system is coupled to an enzymatic DNA construct (DNAzyme), which produces a colorimetric readout when the pathogenâ€™s DNA is detected. Rolling circle amplification is used to ensure an isothermal and sensitive reaction that does not require temperature control (in contrast to current DNA detection methods).

With this detection scheme, a simple color change, visible to the naked eye, will be produced when a pathogenâ€™s DNA is present in a sample â€“ making the test useful for resource-limited settings. The approach is broadly applicable and will enable self-diagnosis of a variety of pathogen-carried diseases including neglected tropical diseases such as leishmaniasis. Direct DNA detection is more reliable than immunological (antigen-antibody based) point-of-care tests as the effectiveness of the antibodies differ between populations, and antibodies persist after treatment, prevent the use of the current immunological tests for test-of-cure.

By adapting state-of-the-art technology into simple diagnostic tools, we are thus addressing one of the most pressing global health challenges that will benefit those who need it the most.

AUTHOR INDEX